Ocean Wealth
POLICY AND POTENTIAL

Ocean Wealth
POLICY AND POTENTIAL

George A. Doumani

Science Policy Research Division
Congressional Research Service
Library of Congress

 **Spartan Books**

HAYDEN BOOK COMPANY, INC.
Rochelle Park, New Jersey

ISBN 0-87671-709-1
Library of Congress Catalog Card Number 72-87830
Copyright © 1973

Printed in the United States of America

Spartan Books are distributed throughout the world by Hayden Book Company, Inc., 50 Essex Street, Rochelle Park, N.J. 07662, and its agents.

1 2 3 4 5 6 7 8 9   PRINTING

73 74 75 76 77 78   YEAR

Foreword

While the oceans have long influenced the destiny of nations and of human culture, only recently has there been a consciously sought nexus with the sea. A marked transition has evolved from random exploration and uncritical exploitation of marine resources to a thoughtful appraisal of bounty from the oceans, and to conflict over sovereignty, extraction, and equitable distribution of these projected benefits. Five unrelated circumstances set the stage for this spectacular change. First, scientific oceanography afforded data for a deeper comprehension of what is in and under the sea. Second, technology developed to permit activities that were historically thwarted by the hostile and strenuous marine environment. Third, as world populations outraced their food supply and terrestrial sources of energy and minerals, the oceans beckoned as a new frontier. Fourth, as these populations industrialized and clustered along coastlines, their mounting waste was indiscriminately dumped into the convenient and seemingly invulnerable sea. Finally, a growing number of coastal and landlocked nations newly granted independence sought their share of benefits from fishing, undersea petroleum and gas, and manganese nodules spread widely on the ocean floor.

As early as 1958, an unprecedented set of treaties on Law of the Sea was spun from a careful eight-year study of these issues. By 1966, however, the President of the United States of America felt obliged to issue a warning about imminent colonialism on the seabed. Major questions of serious portent for all citizens on the planet were examined by President Johnson, the Congress, by officials of other lands, and by the United Nations. New issues arose over public order of the oceans, stimulated by proposals for vesting control over the seabed with the United Nations or a subsidiary agency in efforts to employ the seas peacefully in the interests of all mankind. Geopolitical implications thus involved a unique blend of scientific fact, engineering, and economic analysis with legal and political factors that represented marine extensions of traditional concepts in international diplomacy.

A new phase of inquiry arose with marine-related problems placed on the agenda of the United Nations. The functional, ecological, and legal interconnections, and the awareness that ocean space was of global proportions now led to a systemic rather than fragmented approach to ocean problems. The new concern for the oceans as a common heritage of mankind was heavily tinctured, however, with national self-interest—reflected in positions on boundaries of national sovereignty. Indeed, at the United Nations the issues were considered of sufficient priority that first an *ad hoc* and then a standing committee on the seabed was created as a forum for international debate about the oceans. Subsequently the United Nations set about updating the 1958 treaties, involving fisheries, minerals exploitation, freedom of navigation, freedom for research, protection of environmental quality, and seabed arms control.

During this entire interval, the issue of ocean management was under study in both executive and legislative branches of the United States government. Positions were developed, refined, revised, advocated, and negotiated. Consensus was difficult to generate because the issues were so complex and various public and private interests were in conflict over policies each preferred; the force of tradition remained strong, and enunciation of long-range principles was lost in the noise level of debate.

That debate continues. In addition to being the focus of recent international attention on the sea, the issue of control of the seabed provides a very instructive example of the difficulties inherent in the formation of national policies that necessarily involve combined considerations of science and technology with sensitive economic and political realities.

George Doumani has written an unusually perceptive case study of the evolution of the seabed controversy. *Ocean Wealth: Policy and Potential* portrays the key issues and provides the general as well as specialized reader with a clear objective background to the issues now reaching a crescendo of national and international inquiry.

Beginning with a concise review of geographical and legal definitions of the continental shelf, Mr. Doumani utilizes his professional background as a geologist to explain what potential seabed resources are present and where they are likely to be found. Since knowledge of the occurrence of certain minerals associated with the seabed must be evaluated in the light of prevailing conditions of technology and market economics, Mr. Doumani analyzes the resource potential of each seabed mineral with foreseeable expectations of technological development and economic demand. Such a review of the complexities of resource occurrence for the nontechnical reader is a difficult task to perform without the use of superficial or trivial generalities. Yet Mr. Doumani has presented a clear and informative explanation of the important substance of this technical aspect of the issue.

With a greater appreciation for the resource potential of the continental shelf and the seabed, the reader is next introduced to the still evolving positions of the United States government with regard to the seabed, and how our national policies were modified in response to developments at the United Nations. Mr. Doumani explicitly identifies the key roles of members of the legislature and executive branches in the development of policy positions on seabed disarmament and the proposed seabed regime. For additional reference, Mr. Doumani describes various institutional components of the United States government and the United Nations system that have jurisdictions over the issues under debate.

In an analytical summary, Mr. Doumani offers many insights into the seabed issue, as well as some interesting observations about the increasingly important roles of scientists and engineers in the process of policy formation and review, and the need for development and encouragement of a new breed of scientist—or engineer—policy analyst, or diplomat.

Mr. Doumani has performed an exceptionally useful service to those concerned about the resolution of the seabed debate. By exhibiting a scholarly command of the main elements of marine science and technology and the subtle economic and political realities of policy formation in the United States and in the United Nations, Mr. Doumani has merged and clarified the main points of the seabed issue. All concerned will have a more factual and realistic base of knowledge upon which the critically important decisions about the international role in seabed management must be made.

EDWARD WENK, JR.
Professor of Engineering and Public Affairs
University of Washington

Preface

I must go down to the sea again,
To the lonely sea and the sky,
And all I ask is an oil rig, an echo-sounder,
Some drill pipe, a license to explore,
Some straw bales,
And a slick to steer 'em by.
*—with apologies to John Masefield**

As man's activities progress into the ocean frontier at such rapid pace, national and international concern is increasing over the fate of marine resources and the ocean environment. National jurisdictions over ocean waters and ocean resources have never been clearly defined, and th United Nations has called for a Conference on the Law of the Sea to revise and update the Geneva Conventions of 1958.

The concern of the world community over marine affairs may be summarized by a few basic questions: What is the seabed? What resources are there on and under the seabed? What value do they have? What are the present and future technologies for exploiting them? And what exactly are the nations arguing about?

In the course of my daily work—assisting various committees and members of Congress in legislation for marine affairs—I developed an acute awareness of the need for an authoritative reference book in this area—a text at once nontechnical, concise, and informative. Despite the prevalence of pertinent literature, the bulk of the information is either highly technical, narrowly specialized, or widely scattered.

Ocean Wealth: Policy and Potential is designed to satisfy this need and answer the questions noted. It traces the development in the definition of the seabed and the international conventions that relate to this definition. In dollars and cents, and tons and barrels, it offers inventory of the most significant of the seabed resources, present and potential. Present technological capabilities in utilizing seabed resources, commercially and

---

* Gordon R. S. Hawkins, "Science and Political Will at Sea," *International Journal*, Autumn 1970, p. 704.

militarily, are also discussed with a look at future technological break-throughs, particularly in offshore oil and hard minerals on the deep ocean floor.

The political aspects of ocean affairs are then explored, including the policymaking apparatus in the United States and the forces and events that have influenced development of our policies. From the national scene, the book moves into the international arena and the interest of the whole world about ocean space. It defines the issues that concern all nations over the economic and military effects of seabed exploitation, and looks into problems of managing its resources for the mutual benefit of all mankind. The text concludes with an objective assessment of these problems, in the hope of increasing man's understanding of the proper relationship between scientist and statesman in their common effort toward a world of peace and plenty.

Ocean affairs normally span a wide spectrum of disciplines and subjects, but the problems confronted by the maritime nations are inter-related. The broad coverage of this book, albeit concise, was designed for government officials, political scientists, science policymakers, marine scientists, officials and researchers of the petroleum and mining industries, and all those concerned with energy, the marine environment, and international affairs.

Although the information contained herein is intended for professionals and specialists in the field, the language, style, and format deliberately avoid technical jargon in order to facilitate understanding by laymen interested in ocean science and technology.

Supporting the text are several appendices bringing together, for the first time, complete documentation of official United States presidential orders and legislative acts of Congress, as well as United Nations resolutions. Until the international community agrees to a new law of sea, many years hence, it is hoped that this book will serve as a reference to the resources of the seabed, the ocean policymaking process in the United States, and the marine-related activities of the United Nations.

GEORGE A. DOUMANI

Washington, D.C.

Acknowledgments

The original version of this book appeared under the title *Science, Technology, and American Diplomacy: Exploiting the Resources of the Seabed.* It was prepared as a committee print for the Subcommittee on National Security Policy and Scientific Developments, House Committee on Foreign Affairs, United States Congress. Permission to reproduce it was obtained from Congressman Clement J. Zablocki, Chairman of the Subcommittee.

The author is grateful to his colleagues at the Science Policy Research Division and the Foreign Affairs Division of the Congressional Research Service, who reviewed the original manuscript. Special appreciation is due to Dr. Franklin Pierce Huddle, who shouldered the responsibility for directing the project.

Contents

Ocean Wealth
POLICY AND POTENTIAL

1
Ocean Space

The purpose of this study is to describe the seabed—its configuration and resources—and to show what impact various technological advances to exploit the resources under the oceans have had on national policy and international diplomacy.

AN OVERVIEW OF THE GEOGRAPHY OF THE SEABED

The world oceans occupy more than 70 percent of the surface of the earth. Although the oceans have been divided into Arctic, Atlantic, Indian, Pacific, and Antarctic, this division reflects only the point of view of humans inhabiting the land of the planet Earth. Viewed from the Moon, Earth is essentially a water planet—one large ocean interspersed with continental land masses.

Geologically, the picture is even more radical. The oceans are merely a film of water covering a major portion of the Earth's crust. Other portions of the crust protrude above this film of water and are called land; what is below the water is the seabed.

By virtue of its global characteristic, therefore, ocean space is a common link among land masses, shared by the nations touching this ocean space. Its waters wash indiscriminately the shores of these nations, and its marine life forms journey freely through their grazing grounds heedless of national boundaries. Despite these natural characteristics, ocean space has been zoned off, and national boundaries and jurisdictons have been established by the coastal states.

Until recently, man's needs for the ocean were for the most part confined to food and commerce, followed by military uses. The main concern of nations was the protection of their near-shore areas for their food supply and their commercial fleets. The ocean floor and the seabed were virtually unknown, and their potential resources unheard of.

Progress in marine technology and the widening horizons of scientific inquiry enlarged the sphere of man's knowledge and revealed the presence of natural resources, not only in sea water itself, but also on the ocean floor and in the underlying layers. Today, the sea floor is no longer a bottomless basin but an underwater world with a "landscape" not very unlike man's own world on land. It has valleys and mountain ranges, seamounts and volcanoes, canyons and deep trenches, and a continental margin extending from land to the abyss, all complete with plant and animal life. It is a whole new world, heretofore alien and hostile to man, yet virtually at his doorstep.

Into this underwater realm man has begun to direct his energy—his quest for knowledge, for profit, and for his ultimate survival.

SCOPE AND LIMITATIONS OF THE STUDY

This study defines the area under consideration, the resources of the seabed, and the international activities toward an orderly exploitation of these resources. It presents the geographical and legal definitions of the continental shelf and the sea floor beyond, and the historical background leading to the international concern and the Geneva Conventions of 1958. An inventory is taken of all the resources of the seabed, which are the object of concern among the nations of the world. The development in the techniques of exploiting the seabed are reviewed, showing the present state of the art and what the future holds for underwater exploitation. The economic factors are added to the technological capabilities, in order to assess the parameters interacting in the formulation of policy for exploiting seabed resources.

On the international scene, the United Nations activities are reviewed, particularly following the Malta proposal for an international regime for the seabed. The participation of the United States in these activities is discussed, including United States policy apparatus and the evolution of that policy in international ocean affairs. The role of science and technology is analyzed, showing the effect of technological development on ocean strategy, and the role played by scientists in the diplomatic and policy-making processes.

The study is mainly addressed to the seabed portion of ocean space beyond national jurisdiction. It includes only cursory mention of fisheries and other ocean resources, and the issues of territorial limits.

2

Geography and Legal Concepts
of the Continental Shelf

The crust of the earth as a whole has two major features—the continental platforms and the ocean basins. The physiographic features of the oceans are not merely expressions of the earth's surface but, more significantly, they are reflections of fundamental geological and geophysical provinces of the Earth's crust. These provinces differ in shape, mass, structure, physical and chemical properties, and the composition of their rock constituents.

Where water meets land is not exactly where the ocean basins meet the continental platforms. A relatively narrow margin of each platform is

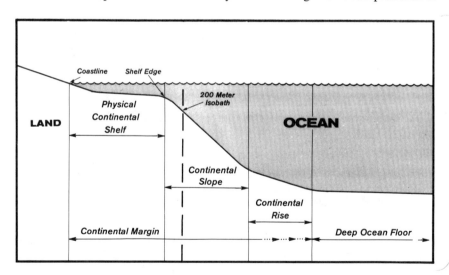

Figure 1. Sketch profile showing the components and average depths of the continental margin. The profile is not to scale; scale has been exaggerated for more contrast, particularly the continental slope. Physically the average width of the shelf is 65 kilometers; that of the slope 15 to 18 kilometers. The average slope of the shelf is 1 degree, that of the slope 4 degrees.

under water, belonging geologically to the continent and not to the ocean basin. This feature is called the continental margin.

The continental margin has three physiographic features: the shelf, the slope, and the rise (Figure 1). The shelf is the extension of the land mass; the slope is its frontal edge; and the rise is that vaguely definable area where the bottom of the slope meets the deep ocean basin.

As its name implies, the continental shelf is topographically a gently sloping terrace, ranging in depth from the mean water line at the shore to a maximum of 300 fathoms* where the sharp slope begins. Where this break occurs, the average depth throughout the world is about 72 fathoms, though for convenience the 100-fathom depth has been adopted.

The width of the continental shelf varies widely from a very narrow shelf off the west coasts of North and South America, to hundreds of miles along Arctic Europe and the Far East. It ranges from a minimum of less than 1 mile to a maximum of 800 miles. Although the depth of the shelf has been used as an international limit, it is the width that determines the area of the shelf and, hence, its significance for the exploitation of its resources.

CONTINENTAL SHELF OF THE UNITED STATES

The United States of America has a coastline approximately 12,000 miles long, with a continental shelf of 650,000 square miles at the 100-fathom depth, including the East Coast, West Coast, and Alaska.

Off Newfoundland, the shelf width increases greatly, averaging over 200 miles. To the south and east are the Grand Banks, which average only about 30 fathoms in depth and stretch eastward for more than 450 miles. If the shelf were limited to a 100-fathom depth, the width would be only about 200 miles.

* The International Committee on the Nomenclature of Ocean Bottom Features proposed the following definition. *Continental shelf, shelf edge, and borderland*: The zone around the continent, extending from the low-water line to the depth at which there is a marked increase of slope to greater depth. Where this increase occurs the term "shelf edge" is appropriate. Conventionally, the edge is taken at 100 fathoms (or 200 meters) but instances are known where the increase of slope occurs at more than 200 or less than 65 fathoms. Where the zone below the low-water line is highly irregular and includes depths well in excess of those typical of continental shelves, the term "continental borderland" is appropriate.

The same definition was used by a group of marine geologists who chose the depth of 300 fathoms arbitrarily (André Guilcher and others) in preparing a report for the United Nations Educational, Scientific and Cultural Organization, Conference on the Law of the Sea, 1957, 13/2.

One fathom equals 6 feet or 1.83 meters.

TABLE 1. Depth Zones of the Oceans

[Areas in millions of square miles]

		Under 200 Meters		*200 to 1,000 Meters*		*1,000 to 2,000 Meters*	
	Total area	*Area*	*Per-cent*	*Area*	*Per-cent*	*Area*	*Per-cent*
All oceans and seas	105.569	7.909	7.49	4.669	4.42	4.630	4.38
Pacific Ocean plus seas	52.880	2.954	5.59	1.791	3.39	2.084	3.96
Pacific Ocean alone [1]	48.476	.791	1.63	1.252	2.58	1.576	3.25
Asiatic Mediterranean [2]	2.648	1.375	51.91	.245	9.26	.276	10.43
Bering Sea	.659	.306	46.44	.039	5.98	.050	7.72
Sea of Okhotsk	.406	.107	26.48	.160	39.48	.091	22.38
Yellow and East China seas	.351	.285	81.31	.040	11.43	.021	5.97
Sea of Japan	.295	.069	23.50	.045	15.18	.058	19.65
Gulf of California	.045	.021	47.71	.009	20.85	.012	25.89
Atlantic Ocean plus seas	27.502	2.383	8.69	1.624	5.92	1.418	5.16
Atlantic Ocean alone [3]	25.240	1.773	7.03	1.305	5.17	1.084	4.30
American Mediterranean [4]	1.271	.298	23.44	.136	10.67	.172	13.52
Mediterranean Sea [5]	.732	.150	20.44	.165	22.48	.127	17.41
Black Sea	.148	.052	34.97	.019	12.59	.034	23.08
Baltic Sea	.111	.111	99.83	.003	.17	-----	-----
Indian Ocean plus seas	21.613	.889	4.10	.632	2.92	.786	3.64
Indian Ocean alone	21.411	.765	3.57	.575	2.69	.766	3.58
Red Sea	.132	.055	41.45	.057	43.06	.020	14.92
Persian Gulf	.069	.069	All	-----	-----	-----	-----
Arctic Ocean plus seas	3.574	1.683	47.10	.623	17.45	.333	9.34
Arctic Ocean alone [6]	2.766	1.125	40.67	.458	16.54	2.82	10.21
Arctic Mediterranean [7]	.808	.558	69.01	.165	20.45	.051	6.27

[1] Pacific Ocean includes Bass Strait.
[2] Asiatic Mediterranean includes Andaman Sea, South China Sea, Java Sea, Celebes Sea and Arafura Sea.
[3] Atlantic Ocean includes North Sea, Greenland Sea, Norwegian Sea, Kattegat and Gulf of St. Lawrence.
[4] American Mediterranean includes Gulf of Mexico and Caribbean Sea.
[5] Mediterranean Sea includes Sea of Marmara.
[6] Arctic Ocean includes only North Polar Basin and Barents Sea.
[7] Arctic Mediterranean includes Hudson Bay, Baffin Bay and Canadian Straits Sea.

Source: L. R. Heselton, Jr., "The Continental Shelf," Institute of Naval Studies, *CAN research contribution No. 106*, December 1968, p. 8.

Between Newfoundland and Cape Hatteras, the shelf decreases in depth from 80 to 30 fathoms. The channel into the Gulf of St. Lawrence is more than 30 miles wide, and the shelf width varies from about 120 miles off Nova Scotia to less than 20 miles off Cape Hatteras.

From Cape Hatteras south, the shelf gradually widens from less than 20 miles to a maximum of 70 miles off Georgia, then virtually disappears off south Florida. If the Blake Plateau is considered as a portion of the shelf, the maximum width would increase to about 300 miles. This would increase the area by about 50,000 square miles, most of which is at depths of between 300 and 500 fathoms.

In the Gulf of Mexico, the shelf rarely exceeds 100 fathoms in depth. To the west of the Mississippi River the edge of the shelf is about 100 fathoms up to 120 miles offshore. The overall U.S. portion of the Gulf contains about 135,000 square miles of shelf of less than 100 fathoms, of which only 8,000 miles is within territorial waters.

On the west coast, the apparent shelf off southern California is about 10 miles wide with an edge at about 50 fathoms. However, the bottom is irregular, and there are shoals and rises beyond 100 miles offshore which geologically should be considered part of the continental shelf. The true shelf—as opposed to the legal shelf—appears to terminate beyond 500 fathoms in many instances off southern California, and in the southern portion is as much as 150 miles offshore.

For the remainder of North America, there would be little effective change in shelf area by assigning an outer shelf limit greater than 100 fathoms. The shelf of the Bering Sea is very flat and has a pronounced edge at around 70 fathoms, attaining a maximum width of 400 miles.

CONTINENTAL SHELF OF THE SOVIET UNION

The Soviet Union has a coastline approximately 23,000 miles long, with a continental shelf exceeding 1 million square miles up to the 100-fathom depth. The coastline stretches along the Arctic Ocean from Norway to Alaska, and southward along the Pacific Ocean from the Bering Sea to the Sea of Japan.

The shelf bordering the Arctic is not of uniform extent, being several miles wider off the Eurasian coast than off that of North America. North of Norway and adjacent to Russia, the Barents Sea forms one of the widest shelves in the world and also one of the deepest. Off Norway, the 100-fathom line is reached almost immediately offshore while to the east, toward Russian waters, it is as much as 150 miles from land.

The Kara Sea, 250,000 square miles in area, is entirely on the continental shelf. It is mostly of depths less than 100 fathoms, with isolated troughs of about 200 fathoms.

TABLE 2.

Countries with Extensive Ocean Area at Depths Less Than 1,000 Fathoms

Country	Approximate coastline (nautical miles)[1]	Approximate area (square nautical miles) at less than		
		100 fathoms	100 to 500 fathoms	500 to 1,000 fathoms
Argentina	2,100	250,000	25,000	15,000
Australia (including New Guinea)	17,500	625,000	170,000	250,000
(Indian Ocean islands)	——	2,000	20,000	65,000
Bahamas (U.K.)	1,400	37,000	5,000	14,000
Brazil	3,700	200,000	33,000	35,000
Burma	1,230	63,000	10,000	10,000
Canada	11,000	>700,000	>200,000	>100,000
China	3,500	200,000	20,000	10,000
Faeroe Islands (Denmark)	155	6,000	30,000	15,000
France	1,375	41,000	5,000	4,500
(Indian Ocean islands)	——	18,000	43,000	63,000
(Pacific Ocean islands)	——	29,000	40,000	75,000
Greenland (Denmark)	5,000	60,000	200,000	50,000
Iceland	1,080	22,000	40,000	>75,000
India	2,750	80,000	20,000	30,000
Indonesia	20,000	380,000	——	——
Ireland	660	36,000	15,000	7,000
Malaysia	1,850	125,000	——	——
Mexico	5,000	100,000	25,000	25,000
New Zealand	2,770	60,000	225,000	175,000
Norway	1,650	30,000	80,000	35,000
Portugal dependencies	——	60,000	45,000	95,000
South Africa	1,430	46,000	44,000	33,000
(South West Africa)	780	20,000	35,000	10,000
South Vietnam	865	84,000	17,000	25,000
Spain	1,500	20,000	23,000	23,000
(Atlantic dependencies)	——	24,000	11,000	15,000
Thailand	1,300	75,000	15,000	——
U.S.S.R.	23,000	>1,000,000	>400,000	>300,000
United Kingdom	2,800	40,000	25,000	50,000
(Falkland Island and dependencies)	——	30,000	65,000	40,000
(Indian Ocean islands)	——	48,000	20,000	35,000
(Pacific Ocean islands)	——	17,000	17,000	35,000
United States	11,650	650,000	>150,000	>150,000
Venezuela	1,000	27,000	10,000	20,000

[1] Coastlines from *U.S. Department of State Geographic Bulletin No. 3*, April 1965.

Source: Adapted from *Ibid.*, p. 9.

To the east, off Siberia, the shelf edge is generally at about 40 fathoms, and reaches a maximum width of about 400 miles. A combination of the Chukchi shelf and the Bering shelf is often considered the world's widest shelf, extending over 1,000 miles north and south. The shelf narrows to less than 50 miles in width north of Alaska, with the edge still at about 40 fathoms and with depths increasing rapidly beyond.

In the Sea of Okhotsk, the 100-fathom line varies from 20 to 100 miles offshore, and encompasses about one-fourth of the whole area. Except for a deep basin near the Kuriles, all of the sea is less than 1,000 fathoms, an area of some 400,000 square miles.

DIFFERING DOCTRINES OF THE "LEGAL SHELF"

The historical and conventional territorial limits of the coastal nations have long been a subject of international controversy. Even in the United States, controversy and litigation were carried on, particularly between the individual states and the Federal Government. Since this paper is concerned mainly with the continental shelf, the near-shore boundaries will be discussed only as they pertain to the subject. Most of the maritime nations of the world recognize and claim 3 nautical miles as the territorial sea, with a 9-mile contiguous zone beyond that.* The rest of the 109 sovereign states that border the sea claim a wider territorial sea which may be as much as 200 miles offshore, as is the case with Argentina, Brazil, Ecuador, El Salvador, Korea, Nicaragua, Panama, Peru, and Uruguay (Appendix 1).

Where the 200-mile figure originated is not very clear. But during a Senate floor debate on the unlawful seizure of U.S. fishing vessels off the South American shores, the late Senator Bartlett asked if those countries had in fact established a 200-mile territorial sea limit. Senator Warren Magnuson answered:

Yes. Now I have a strange anecdote to relate about the 200-mile limit. In Peru, I held talks with the highest officials of the government about the 200-mile limit. They looked me squarely in the eye and said, "We did not establish the 200 miles. You did"—meaning we, the United States.

I said, "How is that?" They pulled out a musty old order that had been in a drawer—I guess they kept it handy—issued during World War II by President Roosevelt, establishing a 200-mile neutrality zone around the western part of South America as protection. They picked that up and said it should be 200 miles off their coast for fishing and other territorial matters.[1]

* Congress enacted a law in 1966 establishing this fisheries contiguous zone. This is the same zone established by the Geneva Convention on the Territorial Sea and the Contiguous Zone, in 1958, which "may not extend beyond twelve miles . . ." (Article 24.2).

Three South American countries (Chile, Ecuador, Peru) arrived at the 200-mile figure by taking the western limit of what they termed 'bioma.' The delegates of these countries at the Santiago negotiations on fishery conservation problems defined the "bioma" as "the whole of the living communities of a region which, under the influence of the climate and in the course of centuries, becomes constantly more homogeneous, until, in its final phase, it becomes a definite type. . . . The western limit of these 'bioma' are variable, and they are wider opposite the Chilean coast, and narrower opposite Ecuador, but the mean width may be taken to be about 200 miles." [2]

The physical dimensions of the continental shelf have not been used to delineate the extent of jurisdiction of the coastal states for the seabed. One obstacle was the lack of complete and accurate data which could be used by the coastal states throughout the world. A more compelling reason was the absence of uniformity in the widths of the continental shelves. Some nations have hardly any shelf to speak of; the conveniently adopted 100-fathom (200-meter) isobath is within even their territorial seas and contiguous zones.

As the importance of the continental shelf began to increase, some nations undertook, by unilateral action, to establish policy and jurisdiction over their continental shelves. In the United States this action was accomplished through an official proclamation of policy by President Harry S Truman in 1945, subsequently referred to as the Truman Proclamation. This proclamation had the effect of opening a Pandora's box for other nations bordering the seas, regardless of whether they possessed the technological capabilities to utilize the seabed as did the United States.

The decade of the fifties witnessed several attempts to define the continental shelf and the coastal boundaries. The United States Congress passed the Submerged Lands Act of 1953, followed a few months later by the Outer Continental Shelf Lands Act. In 1958, representatives of the world's maritime nations at Geneva produced a multilateral agreement on the law of the sea in what are referred to as the Geneva Conventions.

These events leading to the legal delineations of the continental shelf are discussed below in chronological order.

The Truman Proclamation

On September 28, 1945, two policy proclamations on ocean affairs were issued by President Harry S Truman. The first (No. 2667; Appendix 2) established a national policy with respect to the natural resources of the subsoil and seabed of the continental shelf; the second (No. 2668; Appendix 3) proclaimed U.S. policy with respect to coastal fisheries in certain areas of the high seas.

In the first proclamation, the Government regarded as "reasonable and just" the exercise of jurisdiction over the natural resources of the subsoil and seabed of the continental shelf by the contiguous nation. It recognized that the continental shelf was to be regarded as an extension of the land mass of the coastal nation and thus naturally "appurtenant" to it:

> . . . the Government of the United States regards the natural resources of the subsoil and seabed of the continental shelf beneath the high seas but contiguous to the coasts of the United States as appertaining to the United States, subject to its jurisdiction and control.

The proclamation did not specifically delineate any boundary lines or numerical extent of the continental shelf. However, a news release issued on the same day by the White House (Appendix 4) explained that this proclamation did not prejudge the question of Federal versus state control, and that it was intended to enable:

> the orderly development of an underwater area 750,000 square miles in extent. Generally, submerged land which is contiguous to the continent and which is covered by no more than 100 fathoms (600 feet) of water is considered as the continental shelf.

In order to differentiate between the seabed and the subsoil on the one hand, and the superjacent water on the other, the second proclamation was issued, declaring that:

> the Government of the United States regards it as proper to establish conservation zones in those areas of the high seas contiguous to the coasts of the United States wherein fishing activities have been or in the future may be developed and maintained on a substantial scale. . . . The United States regards it as proper to establish explicitly bounded conservation zones in which fishing activities shall be subject to the regulation and control of the United States.

The declaration went further in conceding similar prerogatives to all other nations, concerning the "right of any state to establish conservation zones off its shores in accordance with the above principles, . . . provided that corresponding recognition is given to any fishing interests of nationals of the United States which may exist in such areas." The proclamation emphasized that the " . . . character as high seas of the areas in which such conservation zones are established and the right to their free and unimpeded navigation are in no way thus affected."

These two assertions seemed to imply that all nations should come to an understanding, negotiate fisheries treaties, and respect agreements concerning fisheries regulation and conservation on the high seas. Furthermore, in context with the events of the time and despite the disclaimer in the White

House press release, the continental shelf proclamation might have been an expression of White House strategy for the claims of the Federal Government to the right over offshore oil reserves, the so-called "tideland" disputes. Unfortunately, the two Presidential proclamations led to widely varied interpretations, internationally and domestically, despite the fact that the proclamations did not legally alter the 3-mile territorial limits of the United States.

Within a few years of these presidential actions, numerous coastal nations issued similar proclamations, but without distinguishing systematically among fishing zones, the seabed and subsoil of the continental shelf, and the concept of the high seas. Although these proclamations varied in scope, they included rights which were not then considered within the acceptable regulations of the international community. For example, Mexico and Argentina claimed jurisdiction over their respective continental shelves, including fisheries, but without interference with free navigation on the high seas. Other South American nations went even further, claiming rights over the shelf, the water above it, and the air space above.[3] Nations having very narrow shelves simply extended their claims of exclusive sovereignty and jurisdiction 200 miles offshore, to include the seabed and the subsoil and fishing rights.

In the United States, the legal principle of multiple use of a resource applies to public lands and the navigable waters of streams. The Truman Proclamation in essence extended this legal principle into the sea. Imparting to it another dimension, the proclamation established a distinction between the use of the seabed and that of the overlying water. It asserted the basic premise that each nation possesses sovereign rights over the exploration and exploitation of the natural resources of its continental shelves.

Some analysts view the proclamation as detrimental to international relations and the interests of the United States. They contend that those not acquainted with the national interests of the United States in the proper context of international relations as a whole tend to consider it in the U.S. interest to establish boundaries as far out into the ocean as possible and establish exclusive jurisdiction over everything within them. For example:

> The trouble with this parochial view is that whatever the United States can do in this respect it has to agree that other countries can do the same thing. The reaction we got from the blunder of issuing the Truman Proclamation on Fisheries in September, 1945, is that other countries will claim more than any new claim the United States makes, deliberately interpret the new claim the United States makes in their favor, and use our new claim, their new claim, and their misinterpretation of our new claim, as substantiation for any action they wish to take over and above what the United States wants to do. The parochial view noted . . . above pushed us into this invidious position in 1945, and we should guard carefully against repeating that mistake.[4]

Regardless of the diverse reactions and interpretations, the Truman Proclamation remained de facto policy for many years to come. In the ensuing years, the lack of definitive boundaries and agreement on such boundaries resulted in a series of spectacular cases between the Federal Government and the states, particularly California, Texas, and Louisiana. These litigations concerned the coastal zone, but included areas within the boundaries of the continental shelf.

The Submerged Lands Act of 1953

In an effort to resolve the issue of state boundaries, the Federal Government instituted an action in the United States Supreme Court against the states of California, Texas and Louisiana. These states were chosen because they were then the only states in the Union which had offshore areas with promising oil and gas deposits.*

Between 1947 and 1950, the Court had decided that these states had no title to, or property interest in, the submerged lands off their respective coasts outside the inland waters. The Federal Government claimed all rights over the lands, minerals and other things underlying the offshore waters.

These Supreme Court decisions were reversed by the Congress in the Submerged Lands Act of May 22, 1953.†

The Submerged Lands Act attempted to define certain terms and solve some of the problems, providing definitions of "coastline," "land beneath navigable waters," and the seaward boundaries of the states (Appendix 5). It set the seaward limit as 3 marine miles from the coast, or as these boundaries "existed at the time such State became a member of the Union, . . . but in no event shall the term 'boundaries' or the term 'lands beneath navigable waters' be interpreted as extending from the coastline more than 3 geographical miles into the Atlantic Ocean or the Pacific Ocean, or more than 3 marine leagues into the Gulf of Mexico."‡

The act left undefined the term "inland waters" and introduced a new element of uncertainty as to the historical boundaries of the states. It did,

* California claimed 3 marine miles, Louisiana 27 marine miles, and Texas to the outer edge of the continental shelf. This litigation was referred to as the "tidelands controversy," although in the technical sense neither the tidelands nor the inland waters was at issue. The Federal Government had already conceded the ownership of these areas by the states.

† Article 4, Section 3, of the Constitution of the United States vests in Congress the power to dispose of property belonging to the United States. The power of Congress to grant submerged lands to the states as it did in the Submerged Lands Act of 1953 was challenged the year after, but the act was sustained. *Alabama vs. Texas* (1954), 347 U.S. 272.

‡ One marine league equals 2.4–4.6 miles. Three marine leagues in the sense above equal approximately 10.5 miles.

however, render considerable statutory weight for the first time to the Truman Proclamation by providing that the natural resources of the continental shelf seaward of the areas granted the states "appertain to the United States, and the jurisdiction and control of which by the United States is hereby confirmed." Despite this confirmation, the lack of definition and the uncertainty about the historical boundaries of the states once again led to litigation.

Relative to the continental shelf beyond the undefined inland waters, that is, beyond the seaward boundaries, the Submerged Lands Act also omitted any provision for the administration of the seabed and the subsoil and the natural resources thereof.

On the same day of the issuance of the Truman Proclamation (September 28, 1945), the President issued an Executive order (Appendix 6) reserving and placing certain resources of the continental shelf under the control and jurisdiction of the Secretary of the Interior. A few months before passage of the Submerged Lands Act, on January 16, 1953, President Truman issued another Executive order (Appendix 7) setting aside submerged lands of the continental shelf as a naval petroleum reserve. This action concerned particularly oil and gas, and revoked the former Executive order by transferring the jurisdiction from the Secretary of the Interior to the Secretary of the Navy. The main thrust of the order provided that:

the lands of the continental shelf of the United States and Alaska lying seaward of the line of mean low tide and outside the inland waters and extending to the furthermost limits of the paramount rights, full dominion, and power of the United States over lands of the continental shelf are hereby set aside as a naval petroleum reserve and shall be administered by the Secretary of the Navy.

This Executive order prevailed until revoked a few months later by the Outer Continental Shelf Lands Act.

The Outer Continental Shelf Lands Act of 1953

The Truman Proclamation asserted the rights of the United States on the basis of the geologic unity of the continental shelf with the adjacent continent. Although the proclamation did not establish an official width for the shelf, the accompanying news release set the limit at 100 fathoms (600 feet). These rights were given statutory weight by the Submerged Lands Act of 1953, but the continental shelf as a whole remained vague and undefined.

The Submerged Lands Act defined the width of the shelf to be between the base line at low water and the 3-mile seaward limit of the territorial sea. This zone was therefore considered to be the inner continental shelf. The rest of the continental shelf, seaward of the territorial waters, was then referred to as the *outer* continental shelf.

The Outer Continental Shelf Lands Act was signed as Public Law 212 (Appendix 8) on August 7, 1953. The main thrust of the act was to provide for the administration of the resources of this area. It vested this authority in the Secretary of the Interior, revoking the previous Executive order which had set this area aside as a naval petroleum reserve under the administration of the Secretary of the Navy.

In its declaration of policy the act provided "that the subsoil and seabed of the outer Continental Shelf appertain to the United States and are subject to its jurisdiction, control and power of disposition as provided in this Act." But it made it clear that the act shall not affect "the character as high seas of the waters above the outer Continental Shelf and the right to navigation and fishing therein."

This represented a radical and significant departure from the jurisdictions asserted under the Truman Proclamation and the Submerged Lands Act, in which reference was made only to the "natural resources" of the seabed and subsoil. In the final version of the Outer Continental Shelf Lands Act this phrase was omitted. But despite this omission, the character of the rights claimed remained limited to "jurisdiction, control, and power of disposition."

The act did not define the extent of the outer continental shelf, seaward of the territorial limits. However, in the publication "Description of Outer Continental Shelf," which was part of the legislative history of the statute, the Senate Committee on Interior and Insular Affairs defined the shelf as:

... the extension of the land mass of the continents out under the waters of the ocean to the point where the continental slope leading to the true ocean bottom begins. This point is generally regarded as a depth of approximately 100 fathoms, or 600 feet, more or less. In countries using the metric system, the outer limit of the shelf is generally regarded as a depth of 200 meters, which is approximately the same as the 100-fathom mark adopted by England and America.[5]

In describing the area comprised within these limits, the committee concluded that "the outer shelf can be estimated to contain 261,000 square miles." [6] Computations by the U.S. Coast and Geodetic Survey of inland water areas of the United States, the territorial waters, and the continental shelf are shown below in Tables 3 and 4.

These descriptions, being part of the Senate report accompanying the bill, cannot be considered as having the full stature of the law. They only indicate that the Congress was aware of the geological concept of the continental shelf. Despite this awareness, however, Congress did not adopt that concept when it defined the boundaries of the continental shelf in the letter

TABLE 3. Inland Water Areas of the United States, by Regions
[In square miles]

Locality	Area	Locality	Area
Coastal States		Inland States [1]—*Continued*	
New England	3,149	East South Central	1,116
Middle Atlantic	6,719	West South Central	1,637
Chesapeake	1,688	Mountain	6,936
South Atlantic and Gulf	18,296		
Pacific	19,680	Total, inland	77,127
Total, coastal	49,532	Total, United States	126,659
Inland States [1]		Great Lakes	60,309
East North Central	57,653	Other	66,353
West North Central	9,789		

[1] In general, includes lakes, reservoirs, and ponds having 40 acres or more of area and streams and estuaries, canals, etc., $\frac{1}{8}$ of a statute-mile or more in width. Does not include water surface of the oceans, bays, Gulf of Mexico, Long Island Sound, Puget Sound, and the Straits of Juan de Fuca and Georgia.

Source: National Council on Marine Resources and Engineering Development, *Marine Science Affairs—A Year of Transition.* The first report of the President to the Congress on marine resources and engineering development. Washington, D.C.: U.S. Government Printing Office, February 1967, p. 141.

of the act. Absence of such a definition left flexible the seaward reach of the outer continental shelf; it remained subject to further expansion of United States jurisdiction, either unilaterally or by agreement with other nations.

This flexibility and the absence of any precise definition were exploited in later years in the administration of leases by the Secretary of the Interior. There were occasions when leases far exceeded the 100-fathom depth previously believed to be the intended limit. Accordingly, up to and after the passage of the Outer Continental Shelf Lands Act of 1953, the continental shelf remained undefined.

Geneva Conventions of 1958

Discussion of the territorial sea among the nations of the world dates back to the Hague Codification Conference of 1930, sponsored by the League of Nations.[7] The Preparatory Committee for the Conference devoted considerable attention to the limits of base lines and the widths of territorial waters, but failed to produce the desired convention. Nevertheless, the Committee can be credited with the concept of a contiguous zone, and with focusing attention on the continental shelf and the prospect of

TABLE 4. Area of the United States Continental Shelf, by Coastal Regions
[Thousands of square statute miles]

	Area₁ Measured from Coastline Bounded by—		
	3-Nautical- Mile Band	100-Fathom[2] Contour	1000-Fathom[2] Contour
Atlantic coast	6	140	240
Gulf coast	5	135	210
Pacific coast	4	25	60
Alaska coast	20	550	755
Hawaii	2	10	30
Puerto Rico and Virgin Islands	2	2	7
Total	39	862	1,302

[1] That part of the sea floor extending from the low water line at the coast seaward to the indicator distance or depth.
[2] Fathom is a unit of length equal to 6 feet.

squabbles among the nations over the proper delineation of zones and assertion of rights and jurisdiction.

Following the world's reaction to the Truman Proclamation of 1945, the United Nations made another attempt at codifying the law of the sea. In 1949, the U.N. International Law Commission began a long study across the total spectrum of maritime problems, including the territorial sea, the continental shelf, the high seas, fisheries, conservation, and piracy. These efforts resulted in several draft reports and a final report published in 1956.[8]

The Commission considered that international law did not permit an extension of the territorial sea beyond 12 miles. It also noted that "many States have fixed a breadth greater than 3 miles and [that] many States do not recognize such a breadth when that of their own territorial sea is less." [9] The implication of this observation is that the 3-mile limit is the acceptable conventional breadth, that a "contiguous zone" to 12 miles was within the confines of international law. Although this implication does not constitute a precise definition, the guidelines provided in the Commission report are generally considered the primary basis for recognizing any given breadth of the territorial sea as an international norm.

In 1958, representatives of 86 nations convened in Geneva to participate in the United Nations Conference on the Law of the Sea. They used the reports drafted previously by the International Law Commission as a basis for their deliberations, and the Conference resulted in four conventions (Appendix 9) approved by the U.N. General Assembly:

1. Convention on the Territorial Sea and the Contiguous Zone;
2. Convention on the Continental Shelf;
3. Convention on the High Seas; and

4. Convention on Fishing and Conservation of the Living Resources of the High Sea.

This study deals with only those conventions that reflect upon the zonation and the definition of the continental shelf limits.

The Convention on the Territorial Sea and the Contiguous Zone established criteria for a baseline at the low-water line, the landward side of which is the "inland waters" and the seaward side the territorial sea. The outer limit of the territorial sea was defined descriptively relative to the baseline, but no figures were given to establish its breadth. The contiguous zone was defined as a zone of the high seas contiguous to the territorial sea of a coastal state, where the state may exercise control in such functions as customs, immigration, and sanitary regulation. "The contiguous zone may not extend beyond twelve miles from the baseline from which the breadth of a territorial sea is measured." (Article 24.2)

The Convention on the Continental Shelf (Article 1) defined the shelf as referring "(a) to the seabed and subsoil of the submarine areas adjacent to the coast but outside the area of the territorial sea, to a depth of 200 meters or, beyond that limit, to where the depth of the superjacent waters admits of the exploitation of the natural resources of the said areas. . . ." The natural resources of the continental shelf were defined in Article 2 to include "the mineral and other non-living resources of the seabed and subsoil together with living organisms belonging to sedentary species, that is to say, organisms which, at the harvestable stage, either are immobile on or under the seabed or are unable to move except in constant physical contact with the seabed or subsoil." The problems caused by this definition will be discussed further in the section on the resources of the continental shelf.

DEFICIENCIES OF THE GENEVA CONVENTIONS

The 1958 Conference on the Law of the Sea was followed by another one in 1960. Both failed to delineate the outer limits of the continental shelf. The Convention on the Continental Shelf went into effect in 1964, subject to review and revision in five years after that date. It served to crystallize international law after a fashion, but had three major shortcomings:

First, the convention failed to delineate the territorial sea, which left the matter to the discretion of the individual states and resulted in a few outlandish extensions, hundreds of miles into the sea floor, with exclusive jurisdictions over what is below it, in it, on it, and above it.

Second, the convention ignored the physical characteristics of the continental shelf so that its definition amounted to a legal fiction. The exclusive use of the 100-fathom (200-meter) isobath is arbitrary, scientifically unfounded, and inequitable in the allocation of resources from the sea floor.

It was wise to divorce the superjacent waters from the considerations of sea-bed jurisdiction, but it was unwise to ignore the water surface completely. The nations having a narrow shelf arrive at the 100-fathom isobath almost within eye view from their shorelines. The addition of a lateral extent on the water surface, a specified distance in miles from shore, would have been a fair alternative to the 100-fathom isobath. But the convention left seaward limits undefined, and provided no compensation for countries without continental shelves. Conversely, nations with very shallow continental shelves were given jurisdiction that extended hundreds of miles offshore before reaching the 100-fathom isobath. The Persian Gulf, for example, is in its entirety one continental shelf, according to the definition of the convention, and the Arctic shelf off Siberia is close to 700 miles in width.

Third, the convention contained a delinquent ambiguity inherent in the clause appended to the definition of the continental shelf. Article 1(a) defined the shelf as reaching to a depth of 200 meters " . . . or, beyond that limit, to where the depth of the superjacent waters admits of the exploitation of the natural resources of the said areas. . . ."

The timing of the Truman Proclamation coincided with the increasing development in technological capabilities and the feasibility of exploiting the sea floor. It served to point up a significant aspect in the development of national and international law for the sea—the direct and inevitable correlation between the evolution of law and the development in technology in response to the need for exploitation.

In an early draft of the report of the International Law Commission, a similar correlation between the legal definition and technological feasibility was attempted. The principle of depth-by-exploitability, however, would have permitted countries to claim, as continental shelf, lands far beyond the geological shelf. In 1953 the International Law Commission rejected this concept and adopted the 200-meter limit and the exploitability clause, whence evolved the definition adopted by the Geneva Convention. Even then, the state of the art in offshore drilling had exceeded twice the depth of 200 meters.

The exploitability clause can be interpreted in numerous ways, the simplest of which is that the extent of the continental shelf limit is determined by the capability of exploiting its seabed and subsoil. In other words, one can claim what one can reach.

In the United States, for example, the exploitability clause was construed as authorization under the language of the convention. It has facilitated the leasing of offshore areas far in excess of the 200-meter depth off California, and as far from shore as 115 miles off Louisiana. Together with the Outer Continental Shelf Lands Act of 1953, the Geneva Convention also permitted the United States to claim jurisdiction and control over Cortes

Bank.* By ignoring the ocean depths in all these cases exceeding 200 meters, and by relying exclusively on the technological capabilities and the feasibility of exploitation, the United States is, in essence, asserting rights to further expansion into adjacent areas as its technology permits. But how far is "adjacent"?

It is not inconceivable that the lack of limits to the continental shelf as defined in both the Geneva Convention and the Outer Continental Shelf Lands Act has left the door open for a possible future claim that the continental shelf of the United States extends all the way to Hawaii.

It is evident, therefore, that the dependence of the delineation of the continental shelf on the technological feasibility of exploiting it can be used as license for encroachment. It has already led to confusion and may well lead to grievances among the nations of the world. Continued encroachment would weaken the effectiveness of international law.

These are major shortcomings in the legal definition of the continental shelf. There remains the most important question of all: What is the fate of the deep sea beyond the shelf? The outer limit of the shelf itself being non-existent, it is not surprising that the Geneva Conventions have not addressed themselves to what lies beyond. The legal literature is replete with numerous papers and theses on these issues. They point up the need for revision of the Geneva Conventions and for a fresh look at the oceans as a world entity to be shared by the nations of the world in an orderly and equitable manner.

---

* A group of San Diego businessmen intended to build an artificial island by filling on top of Cortes Bank which lies under 2 fathoms of water. This was to become a nation called "Abalonia." Cortes Bank lies about 110 miles off San Diego and 50 miles off San Clemente Island, and is separated from the island's territorial sea by waters reaching a depth of 1400 meters.

Here, the Bank "admits of the exploitation of natural resources" and is adjacent to the United States of America and, therefore, can be considered part of the United States juridical continental shelf. Furthermore, the Outer Continental Shelf Lands Act defines the legal shelf as all submerged lands seaward of the lands granted to the States and "of which the subsoil and seabed appertain to the United States and are subject to its jurisdiction and control." The Act further authorizes the Secretary of the Army to "prevent obstruction to navigation [as to] artificial islands and fixed structures located on the outer Continental Shelf."

Accordingly, the Secretary of the Interior and the Secretary of the Army formally advised the proposed island builders that their work could not be undertaken without the consent of the United States. [43 U.S.C. Sec. 1333(f) (1964).]

3
Seabed Resources

The focus of world attention, and the main object of international concern in ocean affairs, has been the seabed and the subsoil of the ocean floor, particularly their mineral content. The seabed contains a variety of mineral resources ranging from beach sand and gravel, through heavy minerals associated with beach deposits, to surface deposits of manganese and phosphorite, and subsurface petroleum resources.

DEPOSITS ON THE SEABED SURFACE

Building Materials

The most obvious hard deposits are sand and gravel. In terms of tonnage, this important commodity is by far the most extensively mined throughout the world.* It forms the backbone of the construction industry as aggregate and filler material. Smaller percentages go into glassmaking and other relatively minor industries.

Although most of the sand and gravel is mined at or near the beaches, the ever-increasing populations and their concentration in coastal areas create demands that are expected to push this industry into the offshore areas.† Construction being essential to accommodate these demands, the seabed will increase in value for providing the necessary sand and gravel aggregate for the future.

Calcium carbonate shells and sands fragmented from them are also mined for use in the production of portland cement and lime. Oyster shells and other calcareous shell deposits are mined from several coastal locations in the United States and elsewhere in the world.‡

* Current production in the United States exceeds 50 million cubic yards.

† John Schlee of the U.S. Geological Survey estimates that 50,000 square miles of the continental shelf off New England, Long Island, and New Jersey are blanketed with sand and gravel. A large deposit off the New Jersey coast lies in 66 to 132 feet of water and may contain several billion tons of gravel.

‡ Offshore production of oyster shells exceeds 20 million tons annually.

Heavy Minerals

Several minerals occur in association with, or under, the beach sands overlying the bedrock. These minerals, referred to as heavy minerals* are now mainly derived from sources on land. As the rocks on land undergo the relentless battering of wind, rain, ice, and other destructive agents, the rocks yield to the processes of chemical and mechanical weathering. The weathered fragments are transported by streams and wind to their final destination in the sea. There, the winnowing action of the waves serves to concentrate the heavier minerals and metals into profitably minable deposits.

As would be expected, the heavier the mineral, the closer to shore it is deposited. Consequently, these placer deposits are expected to be located on present beaches, on submerged beaches, and a few miles offshore, near their source rocks on land. They usually include the heavier metals like gold, tin, and platinum, and the relatively lighter minerals like diamond, the titanium minerals ilmenite and rutile, the tungsten minerals scheelite and wolframite, the iron ore magnetite, chromite, and zircon.

Although seabed placer deposits are normally not as abundant nor as valuable as their counterparts on land, several exceptions are found around' the world, such as the tin deposits off the coast of Cornwall in England and off the coast of Indochina, the diamond deposits off the coast of southwest Africa, and the zircon sands of New South Wales, Australia.

Phosphorite

Among the mineral deposits on the seabed, phosphorite has been found in extensive areas of the continental margin, and appears to have a promising potential for the mining industry. Phosphate rock is composed mainly of calcium phosphate, and more than half of what is mined on land goes primarily for the manufacture of agricultural fertilizers. A smaller percentage is used in the manufacture of organic and inorganic chemical compounds of phosphorus.

Phosphorite is formed by the precipitation of phosphates from sea water. These dissolved phosphates have probably come from decayed phosphatic matter in regions where sudden and extreme environmental changes result in massive kills of marine organisms. In its circulation through the oceanic domain, the phosphate-rich water passes through regions where little or no detrital sedimentation is taking place. In these regions of

---

* Minerals are separated from a mixture of crushed rock in a liquid called "bromoform." Bromoform has a specific gravity of 2.85; minerals with specific gravity greater than that of bromoform are referred to as heavy minerals.

depositional quiescence the dissolved phosphates begin to precipitate from sea water and accumulate on the ocean floor.*

Submarine phosphates are deposited in the form of irregular accretionary nodules, flat slabs, and coatings on sand grains and rock fragments. They vary in shape and size, some weighing between 150 and 250 pounds. Large areas of the ocean floor are blanketed with nodular phosphorite. Occurrences are known from very near shore to more than 200 miles offshore, in depths ranging from 60 feet to more than 11,000 feet.† Sampling and preliminary exploration around the world indicate the presence of phosphate rock on the east and west coasts of North and South America, off South Africa, northwest Africa, and equatorial west Africa. In places where oceanographic data were successful in locating sea-water upwelling and detecting phosphate-rich waters, the prospects are considered good that phosphate rock is present. Examples of such upwelling areas include northwestern Australia, the Timor Sea, the eastern Arafura Sea, the Coral Sea, and the Tasman Sea.

Although there has been no production of submarine phosphate rock on a commercial scale, some deposits have been surveyed and current investigations reflect the increasing interest of mining companies in these offshore minerals. The most extensive of these investigations have been those made offshore at southern California, U.S.A., and Baja California, Mexico. Phosphorite nodules were first discovered off the coast of southern California in 1937. Extensive sampling and exploration has been conducted since, the most recent being the investigations of the sand deposits offshore Baja California, Mexico, by Bruno d'Anglejan (Ph.D. thesis, 1965) and by the Global Marine Company in 1966.

John L. Mero (1965) estimated that the California deposits cover about 10 percent of the potentially workable area of 36,000 square miles. The concentration averages about 22 pounds of phosphorite nodules for every square foot of sea floor, and the deposits available for mining exceed 1 billion tons.[10] More recently, the California deposits were estimated at 100 million tons classed as known "marginal" resources, more than 1 billion tons as known "submarginal,"‡ and nearly 2 billion tons of inferred but undiscovered resources.[11]

---

* The phenomenon of upwelling of ocean waters appears to be the most accepted theory for the origin of submarine phosphates. The knowledge of the processes forming phosphate rock and its depositional environment is essential to the successful exploration for, and locating, these deposits.

† During the oceanographic expedition of the British corvette H.M.S. *Challenger* (1872–1876), phosphorite nodules were recovered from a depth of 11,400 feet on the Agulhas Bank off South Africa.

‡ McKelvey defined marginal resources as material that might be produced at prices not more than 50 percent higher than those prevailing now, or with comparable

Considering the continental shelves of the world as occupying 10 million square miles, and assuming a 10 percent deposit similar to that of the California shelf, Mero concluded that the continental shelves of the world should contain 300 billion tons of phosphorite. If 10 percent of this amount was economical to mine, the reserves of sea-floor phosphorite would be 30 billion tons. At the rate of present world consumption, this supply would last 1,000 years.[12]

Depending on its grade and phosphate content, phosphate rock currently mined on land ranges in price from $6 to $12 per ton at the mine site. The value of submarine phosphorite relative to land deposits will be considered later in context with technology and economics.

Manganese Nodules

Undersea deposits of manganese and iron oxides precipitate from sea water in much the same way as do phosphorite nodules—the colloidal particles adhering to any grain or rock fragments and growing by accretion, layer over concentric layer, to form an onion-like structure. Each form assumes the shape of its nucleus, sometimes forming crusts on surfaces of submarine rock outcrops or coatings on animal or plant remains. They range in size from 0.5 inch to more than 6 inches, with an overall average of 2 inches.

Manganese nodules vary greatly in composition, from region to region as well as from nodule to nodule. The chemical constituents are mostly oxides of manganese, iron, silicon, and aluminum, with calcium and magnesium salts. Nodules sampled since the *Challenger* expedition have been analyzed extensively, and numerous other elements have been found. Most of the recent investigations, however, have focused the attention of the mining industry on such constituents as cobalt, nickel, and copper, rather than manganese alone.

Although present knowledge of the distribution and extent of manganese nodules in the world ocean is rudimentary, the nodules appear to be almost ubiquitous. Some oceanographic expeditions have dredged manganese nodules at practically every station in the Pacific, Atlantic, and Indian oceans. Manganese nodules have been found even in Lake Michigan. They are known to occur particularly beyond the continental shelf, in the abyssal plains and oceanic deeps such as the Mariana Trench in the Pacific, deeper than 22,000 feet. Extensive research and analyses have been done on

---

advances in technology. Over the longer period and with technological advances, resources recoverable at costs two or three times more than those produced now may have some foreseeable use and prospective value. These were termed submarginal resources.

Figure 2. Manganese nodules on the Indian Ocean floor, with concentration higher than 50 percent. The photo was obtained in August 1971 on cruise 48, station 53, of the National Science Foundation's USNS *Eltanin.* Approximate location: latitude 34° 28'S, longitude 100° 3'E, more than 500 kilometers west of Fremantle, Australia; water depth about 4,800 meters. *NSF Photo from Smithsonian Institution Oceanographic Sorting Center.*

nodules from numerous localities, particularly on their growth rates and concentrations. Accretion rates in the order of 0.01 to 1 millimeter per 1,000 years are considered normal growth in deep waters. In shallower waters, some samples showed accretion rates close to 1 millimeter per year.

Underwater photographs and television tapes of blankets of manganese nodules have been studied, and concentrations in pounds per square foot have been computed (Figures 2 and 3). An American expert has estimated that the Pacific Ocean may contain more than 1,600 billion tons of manganese nodules which are being formed at the rate of about 10 million tons per year. Russian specialists, on the other hand, estimate a total Pacific Ocean surficial tonnage about one-twentieth of this.*

---

* Mero, *op. cit.,* p. 174, estimated the total Pacific Ocean nodules to be 1,656 billion metric tons. N. Zenkevitch and N. S. Skornyakova, *Natura,* 3: 47–50, 1961, obtained an estimate of 90 billion metric tons.

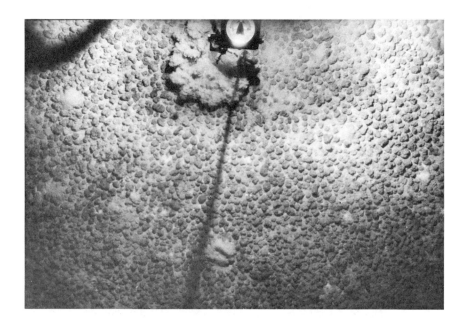

Figure 3. Manganese nodules on the Indian Ocean floor (same location as in Figure 2), with concentration approaching 100 percent. *NSF Photo from Smithsonian Institution Oceanographic Sorting Center.*

Red Sea Geothermal Deposits

Perhaps one of the most significant of recent oceanographic discoveries is that of the bizarre pools of hot brine on the floor of the Red Sea. Unusually saline water in the Red Sea has been known since the Russian expedition of the *Vitiaz* in the 1880s. In the mid 1960s, however, more startling data were obtained by the British RRS *Discovery,* followed by the U.S. (Woods Hole) vessel *Atlantis II,* and several others. The *Discovery* sampled water with a temperature of 111 degrees Fahrenheit and a salinity of 256 parts per 1,000.* *Atlantis II* measured a brine temperature of 133 degrees F (56.5°C) and obtained bottom sediment samples having a temperature of 144 degrees F (62.5°C) and containing a mixture of metal compounds, principally oxides and sulfides of iron, manganese, zinc, and copper. In March 1971, the U.S. research vessel *Chain* revisited the hot-brine area;

---

*Average ocean salinity is $35^0/_{00}$, which means 35 grams of solid material contained in 1,000 grams (1 kilogram) of sea water. Normal Red Sea salinity is $40^0/_{00}$.

new measurements revealed that a body of water having a minimum temperature of 219 degrees F (104°C) has been added since the initial measurement of 1966.*

Geothermal heat from the molten interior of the earth is transmitted to the Red Sea water through the fissures along the rift of the Red Sea floor. The heated waters dissolve salts from sedimentary rock formations and leach heavy metals out of crustal volcanic rocks, creating metal-saturated brine. As this metalliferous brine cools it releases the sulfides of lead and zinc, and the carbonates of iron contained in the water.

This has been the theory generally accepted by oceanographers. The Russian scientist D. D. Kvasov, of the Leningrad Academy of Science, has proposed another theory suggesting that the brine pools may be ancient lakes.[13] Whatever the explanation, the total brine-free sediments at one location only, the Atlantis II Deep, is estimated to be over 50 million tons. It contains appreciable amounts of zinc, copper, lead, silver, and gold which, at current smelter prices, would be worth about $2.5 billion.[14]

The significance of the Red Sea discoveries lies in the interrelated factors of scientific knowledge, exploitation, and legal control. Scientific knowledge will lead to discoveries of similar deposits in submarine environments having analogous geological characteristics. This development, in turn, will have its special economic connotations and present, ultimately, the legal and international problems of exploiting these resources.

Other Surface Deposits

The unconsolidated surface sediments of the ocean floor include several known deposits which could become potentially economical. The pelagic sediments of deep offshore waters are classified as red clays if their organic content is less than 30 percent, and as oozes if their organic content is more than 30 percent. Their widest distribution is in the Pacific Ocean, usually at depths averaging between 6,000 and 17,000 feet.

Although now uneconomical, these clays and oozes contain some potentially useful constituents such as calcium carbonate, alumina, iron oxide, and silica. They literally cover the ocean floor, and are estimated to be in the order of thousands of billions of tons.

Nearer to shore, and in shallower water, unconsolidated green sands (glauconite) are found abundantly in areas of slow detrital deposition. Glauconite sand is a hydrous potassium iron-silicate with a small amount of potash content. It has been sampled off the coasts of Africa, the Americas, Australia, Japan, New Zealand, the Philippines, Portugal, and Britain. Green sand mined from land has been used as a water softener and soil conditioner.

* Ross, David A. "Red Sea Hot-Brine Area: Revisited." *Science*, March 31, 1972, p. 1455.

Barite, a barium sulphate mineral, is another surface deposit which occurs in nodular form like phosphorite and manganese. Barite concretions have been dredged from the continental margin off the coasts of Ceylon, southern California, and the Kai Islands in Indonesia. The concretions range up to 2 pounds in weight and assay around 75 percent barium sulphate.

Large deposits of barite are known on land, which makes the ocean barite unimportant at this time. However, barite is being mined offshore of Barite Island in Alaskan waters 20 to 80 feet deep, with known reserves in 120 feet of water. Current production from this operation alone is 1,000 tons per day;[15] daily production of primary barite in the United States averaged about 2,600 tons for 1970.[16]

DEPOSITS BELOW THE SEABED SURFACE

The subsurface deposits of the seabed are those contained within the structures of consolidated sedimentary and basement rocks. They can be viewed in two groups, related primarily to the methods of their extraction. On the one hand, petroleum, natural gas, and sulphur are extracted through holes drilled into the sea floor. On the other hand, coal, iron ore, and vein deposits are extracted in conventional mining manner by driving shafts and drifts into the seabed from adjacent land areas.

Undersea mines operated from land entries, either from shore or from artificial islands, have been in use for centuries. Magnetic veins are mined near Jussaro Island, Finland, and an extensive undersea iron ore is mined from Bell Island in Newfoundland. Land-entry coal mines are also operated in England, Japan, and Nova Scotia. However, these operations are closely tied to land and do not figure prominently in the future of the seabed.

Petroleum

By far the most important of all marine resources is petroleum. Although land exploration opportunities for petroleum have not been exhausted, petroleum exploration and exploitation have invaded the continental shelves at a rapid pace. Potentially important oil fields are being discovered every year, and the nations of the world are investing considerable capital in offshore ventures.* More than 85 countries are engaged in

* A single lease sale of offshore tracts in the Santa Barbara Channel of California brought the Department of the Interior over $600 million in February 1968, and in March a similar amount was obtained from tracts off the Gulf of Mexico shores. Altogether the revenues for that fiscal year amounted to more than $1.5 billion. On December 15, 1970, the Department of the Interior received some $850 million from oil companies in bids for 127 underwater tracts off the Louisiana coast. A single tract brought a bid of more than $38 million.

offshore activities, and discoveries have been reported from the shelves of North and South America, Australia, Japan, the Mediterranean countries, the Red Sea, the Arabian Gulf, the Union of Soviet Socialist Republics and, most recently, in the North Sea and the South China Sea. Thirty-two of these countries are already producing petroleum from their continental shelves, which accounts for 16 percent of the world's oil and 6 percent of the world's natural gas. It is expected that by 1980 this percentage will double or quadruple.*

The extent of petroleum deposits offshore cannot be determined, and numerous estimates have been advanced. Proved reserves in the "free world" are estimated to exceed 500 billion barrels of oil† and nearly 1.5 *million billion* (quadrillion) cubic feet of gas.[17] It is believed that out to a depth of 1,000 feet, nearly two million square miles of the shelf areas are geologically favorable for petroleum occurrence. It could be safely assumed that nearly every coastal nation has offshore areas that are favorable for petroleum accumulation. The imprecision in the estimates is largely due to the lack of adequate data, and the lack of knowledge necessary for projecting and predicting these estimates.

This knowledge depends basically on understanding the origin of petroleum and the factors required for its accumulation. Several theories exist on the origin of petroleum, but the one that has been most widely accepted holds that the origin of hydrocarbons is organic. The hydrogen and carbon originated from the remains of plant and animal life that existed millions of years ago in former seas or swampy environments. Such life forms were presumed to have been very small, probably microscopic. Support for this theory is derived from interpretation of the geological records, and studies of oil fields and oil-bearing formations that have already been explored and developed throughout the world.

After having been formed, oil accumulates in reservoirs formed by sedimentary layers called formations. Subsequent movements in the earth's crust result in two distinct processes: One is deformation of these layers; the other is their transformation into hard rock.

As the layers are compressed, the oil accumulated in the sediments is forced to migrate into pervious sand bodies with pore spaces between the particles that facilitate the mobility of the oil. Meanwhile, compaction and particle cementation have turned the loose sediments into rocks, the sand becoming sandstone, the silt siltstone, and the mud mudstone or shale.

---

* Industry experts predict that world production of offshore petroleum will exceed 20 million barrels per day, compared with today's 6.5 million. The present U.S.S.R. offshore production exceeds 90 million barrels a year, less than 4 percent of that country's total output. In Moody's September 21, 1970, "Stock Survey" oil and gas produced from offshore wells are predicted to quadruple by 1980.

† Proved crude reserves for the "free world" for the year 1967 were shown to exceed 525 billion barrels. (*Oil and Gas Journal*, May 6, 1968, p. 77.)

While sandstone is the ideal medium to contain the oil, limestone, and other porous rocks also are often oil-bearing. The shale or mudstone is the ideal rock to seal it. Also, in order for the oil to stay in the pervious sandstone to which it has migrated, it has to have an impervious layer over it to check its migration.

The earth's crust is mobile and dynamic, forever on the move. These movements result in deformation which manifests itself in the form of up-lifted mountains, downwarped valleys, and twisted and contorted sedimentary strata. The beneficial part of this upheaval is the creation of structural forms which provide the traps and reservoirs necessary for the containment of oil. The types of traps where oil has been found are numerous; however, one of the most ideal structures is a dome called "anticline" (see sketch A, Figure 4).

In an ideal situation, a reservoir would be a closed sequence of sedimentary rocks, including a layer of oil-bearing sandstone capped with a layer of shale or mudstone. The contents of this reservoir include some water left over from the former seas, the oil body floating over the water, and the natural gas at the very top, all in the pores and cavities of the host rock.

The salt domes of the Gulf of Mexico are cylindrical bodies of salt that have been squeezed from their parent formations, piercing upward through a succession of overlying sedimentary layers (see sketch B, Figure 4). Against the walls of a salt dome the oil-bearing layers are closed off, providing a reservoir and a seal for trapping the oil. Such diapiric structures have been known to occur farther offshore and in deeper waters in the Sigsbee Abyssal Plain beneath the floor of the Gulf of Mexico. Some structures have protruded above the flat surface of the Plain and are termed "knolls." Those without surface expression are "domes." Recent seismic discoveries have revealed the presence of diapiric structures in numerous offshore locations around the world.[18] The Sigsbee Knolls were discovered in 1954 by the *Vema*, the research vessel of Columbia University's Lamont-Doherty Geological Observatory. In 1968, scientists on board the *Glomar Challenger* drilled through the cap rock of the Challenger Knoll in 11,720 feet of water and recovered a core 472 feet long. The sediments in the core contained oil and were similar in composition to other salt domes onshore and offshore.[19]

Except for this type of reservoir, almost all production areas offshore are geologically related to fields onshore. Since the continental margins are essentially submerged edges of the continents, knowledge of petroleum habitats on land can be applied offshore with considerable certainty.

Beyond these near-shore areas, the petroleum potential of the outer continental shelf and slope has been little investigated. But there are indications that the presence of petroleum source beds is very likely in the continental slope, and progressively less beyond the slope into the abyssal plains and oceanic deeps. Exceptions, of course, can be found in depositional

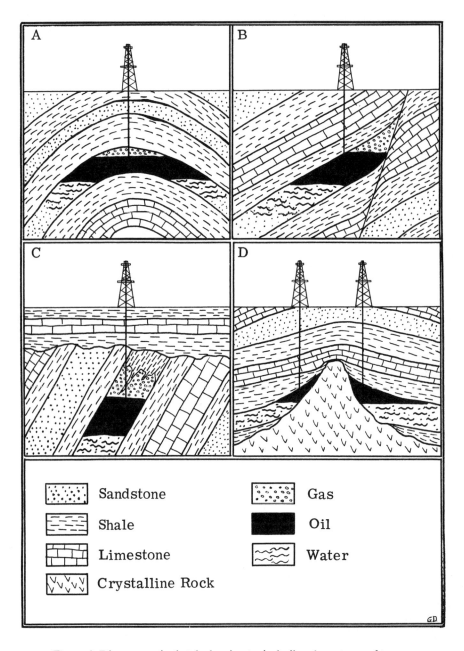

Figure 4. Diagrammatic sketch showing typical oil and gas types of traps.

ocean basins with great sedimentary accumulations and salt-dome structures like those described above. However, as with other ocean resources, the exploration and exploitation of petroleum resources of the seabed depends on the technological capability and economic feasibility for future development.

Other Subsurface Deposits

Deposits of sulphur, coal, salt, and potash are known to occur on the continental shelves of many seas. Sulphur has been mined from salt-dome structures in the Gulf of Mexico, and similar structures are known in the Arabian Gulf, the Red Sea, and the Caspian Sea.

In the Gulf of Mexico, sulphur occurs in considerable quantities in the rock formations capping the salt domes. Several of these domes are now being mined on land, and an elaborate mining operation is being conducted off the coast of Louisiana. In 1968, more than $35 million worth of sulphur came from two mines, one located 7 miles seaward of Grand Isle and the other at the Caminada Pass. The sulphur is extracted by the Frasch method in liquid form through a drill hole similar to an oil well. The most important single sulphur market in the world is the fertilizer industry in Florida, where sulphuric acid is used in processing local phosphate rock into fertilizer. Consumption of elemental sulphur until the early 1960s has increased at an average of 5 percent annually, and jumped to an average of 12 percent annually since then. Present reserves are estimated at nearly 40 million tons, and more sulphur is expected to occur in similar domes offshore.

In the bedrock below the sedimentary cover of the ocean floor, the crystalline rocks, like granite, may contain metallic minerals. Farther offshore into the abyssal plains, the crystalline bedrock is mostly basaltic rather than granitic, where chromite, nickel, and platinum may be found. Unlike the dredging techniques for mining surface deposits, and the conventional land-based mining methods, the extraction of deep-ocean minerals presents formidable problems. Given the present availability of such minerals from land ore deposits, the consideration of deep-ocean minerals becomes a highly academic one of potential rather than actual resource. Nevertheless, the onrush of technology makes it hard to predict whether or when deep submarine deposits may become "ores." They should not be ignored in the formulation of policy for the long-range future.

RESOURCES OF THE OCEANS

Although attention in this report is focused on the seabed, it is appropriate to give some consideration to the resources of the sea itself, to their exploitation by man, and to the potential they offer for the future. They include the living organisms of the marine environment for products like food, food derivatives, and pharmaceuticals; potable water from the sea; the salts and other minerals contained in the water; the minerals on and under the ocean floor; and such related activities as shipping and aquatic recreation.

Food from the Sea

Since his emergence on earth and his first encounter with the sea, man has utilized it as a source of food and a means of transportation.

Life on earth emerged from its primordial ocean, and the life-giving characteristics of the ocean have always played their role in the maintenance of the food chain of living organisms. Man's quest for food from the sea went through the hunting stage for most of historic time, and is now extending to the domestication and farming stages analogous to those known on land. Methods of utilizing the food resources of the sea have also progressed. They vary among the nations of the world from primitive techniques to ultramodern electrical, electronic, and acoustical fishing methods. The immediate results of these technological developments are the obvious overfishing and extermination of a number of species, and the diversification into other types of living organisms that have been heretofore neglected. The world harvest is estimated at 60 million tons of fish annually from conventional species. Estimates of the amount of food that can be harvested from the sea range from 200 million to a potential of 7.2 billion tons, with an acceptable figure of 400 million metric tons.[20]

Fish Protein Concentrate

Technological progress has resulted in methods of utilizing that part of the fish catch that is not considered at this time to have the qualities preferred for the dinner table. So-called trash fish can now be processed into a tasteless, odorless powder called Fish Protein Concentrate. FPC powder is 75 percent protein, one pound of which is equivalent to 5½ pounds of beef or 2½ pounds of nonfat powdered milk. At about 40 cents a pound, it is a cheap source of protein that can be added to staple foods such as bread, pasta, soup, and other foods, to supply the minimum daily requirement of protein. The potential of this product in alleviating protein deficiencies and combating worldwide disease and hunger offers one answer to the problem of feeding the multiplying numbers of the human population.

Aquaculture

The change in man's techniques from hunting the ocean to farming it has progressed in great strides. Although domesticating and herding marine animals, and planting and harvesting marine plants, are not nearly on a scale comparable with such practices on land, aquaculture techniques have been known and practiced for hundreds of years. Recent developments, however, have been evident throughout the world, and aquaculture farms for the production of seafood are being established according to scientific specifications which control the organisms from the breeding stages through marketing and distribution. Aquafarms are not confined totally to estuaries and seashores; many of them are found farther inland. A good example in the United States is to be found in Arkansas where gross sales from aquaculture exceeded $15 million in 1969. Elsewhere, during a recent visit to Lebanon, the author was surprised to learn that the trout served in the restaurants of Beirut is supplied locally from an inland aquafarm. In a most unlikely place near the inland town of Jezzine, he visited a trout farm hardly noticeable to the unsuspecting visitor. The concrete ponds were built on terraces, clinging precariously to the steep slope typical of the rugged Lebanese mountains. The ponds were teeming with healthy trout in all stages of growth, and the whole venture appeared to be a viable and lucrative business.

Aquaculture is by no means confined to finfish. It includes also shellfish like oysters and clams, crustaceans like lobsters and shrimp, and such oddities as turtles and bloodworms. Throughout the world, the volume of the products of aquaculture has grown to a total of 4 million tons. The United Nations Food and Agriculture Organization estimates that by 1985 world aquaculture could expand to 20 million tons.

Drugs from the Sea

In utilizing the living resources of the sea, the extraction of medicinal and pharmaceutical products is often little known and unpublicized. However, the use of ocean organisms for medicinal purposes has been known since the early civilization of the Chinese and in Biblical times. Nowadays, thousands of marine organisms are known to contain biotoxic substances, and yet fewer than 1 percent have been closely examined or thoroughly evaluated for their medicinal characteristics. Marine biomedicine has been gaining prominence, and more attention is being focused on the potential of the ocean environment as a source of new medical discoveries and new drugs.

Sea Water and Its Minerals

The most abundant and most essential of world resources is the water of the ocean. The accelerating increase in human population and the soaring demands for industrial and agricultural use of fresh water have created acute water shortages around the world. The worsening pollution of the freshwater supply and the demands of arid areas where normal water supplies are nonexistent highlight the need for turning to sea water.

Desalination—the production of potable water from sea water—is now an active concern, and many countries suffering from water shortages are erecting and utilizing desalting plants for their domestic needs. More than 600 desalting plants of 25,000 gallons per day capacity or greater are in operation or under construction in the world. Their total production of potable water exceeds 200 million gallons per day. It is estimated that by 1975 worldwide utilization of desalting plants will result in the production of 1 billion gallons of fresh water per day.[21]

The sea holds about 330 million cubic miles of water, which contain an average of 3½ percent of various elements in solution. Each cubic mile weighs some 4.7 billion tons and holds about 166 million tons of solids.[22] The most abundant of these is common salt—sodium chloride. Its two main elements, sodium and chlorine, constitute more than 85 percent of all the solids dissolved in sea water. Salt has been extracted from sea water since time immemorial, and although some of the machines and tools used today are modern, the basic techniques of salt mining by solar evaporation is still prevalent around the world.

Besides common salt, sea water contains commercially extractable amounts of magnesium metal and compounds, and smaller amounts of bromine, iodine, calcium, and potassium compounds. There are at present some 300 near-shore operations in 60 countries engaged in the production of these minerals.[23] Quantitative chemical analyses reveal the presence in sea water of all but about a dozen of the stable elements; advanced analytical techniques may eventually reveal the presence of virtually all known elements in sea water.

One of the latest milestones in mineral extraction from sea water was the invention of a process for extracting uranium oxide. Developed by N. J. Keen at Britain's Atomic Energy Research Establishment at Harwell, the process promises a final product at $20 per pound.

Of the more appealing constituents of sea water, precious metals have been the object of unrewarding research and experimentation for some time. Silver is the most abundant precious metal in sea water, but most attempts have concentrated on the extraction of gold. Numerous gold-extraction processes have been patented, but none has yet been found that could be classified as economical. Gold concentration in sea water has been found to vary from 0.001 milligrams per ton to almost 60 milligrams per ton, with an average about 0.04 milligrams per ton. Of all these attempts, only one case is known where any measurable amount of gold was actually obtained. This was done by the Dow Company in their bromine extraction facility in North Carolina. Fifteen tons of sea water were processed, producing 0.09 milligrams of gold, worth about .01 cent.[24]

Regardless of the number of elements and compounds present in sea water, their extraction depends on a technology that would make the effort economically profitable. At the present level of technological capability, and taking into consideration such factors as operating costs and cheaper, competitive sources, distribution costs, and consumption rate, it has been shown that only six minerals can be profitably extracted. These are: salt, magnesium, sulphur, potassium, bromine, and boron. Any mineral with concentration below that of boron is considered economically unprofitable to extract.[25]

4
Technology and Economics

The discussions thus far have dealt with "deposits" on, in, and under the ocean floor. Under the proper circumstances such deposits are converted by man into "ore bodies." To a geologist, *an ore is defined as a deposit that can be mined at a profit.* In order to accomplish this conversion, a venture requires the quantitative presence of a deposit in a certain environment, and the technology to extract the mineral, process it, and market it at a profit.

A close relationship thus exists between technology and economics, in which the recovery of resources is a business venture primarily economic and only secondarily technologic. The economic factors, however, interact in a complex manner sometimes reflecting the immediate impact of technological development.

In the case of the ocean resources, mining of hard mineral deposits depends mainly on economics. The extraction of subsurface deposits, chiefly petroleum, involves a closed sequence in which economic demands spur technology and technology subsequently pushes the market to demand further technology. This section examines the economics and technology for the extraction of hard minerals, oil, and gas. The most significant of the hard minerals are the phosphorites and the manganese nodules, both of which need to be harvested by mining techniques. Oil and gas resources are extracted through holes drilled in the seabed—a technique already in practice, within water-depth limitations.

HARVESTING HARD MINERALS FROM THE SEABED

Mining hard mineral deposits from the sea floor resembles more closely the harvesting of fish than mining as practiced on land. The equipment used for undersea mining operations is simple and unsophisticated, adapted for the most part from similar land mining equipment. Table 5 shows the types of mining techniques and the timetable projections for the year 2000 to a depth of 1,000 feet.[26] For the immediate future, however, the most promising of these techniques is essentially the principle of dredging. Some of the

TABLE 5.—Ocean Mining Technology Timetable

	Depth of Water			
	50 Feet	*300 Feet*	*600 Feet*	*1000 Feet*
Mining using air-lift device	1960	1970	1975	1980
Mobile miner (ocean floor)	1970	1972	1975	1980
Barge dredge lift	1900	1970	——	——
Stationary mining platform	1960	1970	1975	1980
Buoyant submersible system	——	1975	1977	1980
Underwater open-pit hard-rock mining	1975	1985	1995	2005
Underwater "aerial" photographic reconnaissance	1960	1964	1970	1975
Exploration submarine (corer)	——	1968	1968	1968
Underwater site development station	1970	1972	1975	1980
Solution mining (sulfur, potash)	1961	1980	1985	2000
Hardrock mining (below shelf)	1900	1985	2000	2000
Mining shaft	1970	1980	2000	2000

dredges have the capability to dig and scoop consolidated sediments. Other types employ hydraulic pumping and air-lifting actions, strictly for surface unconsolidated sediments.

Conventional dredging is done by four main types of dredges:

1. Bucket-ladder dredge, limited to 150 feet.
2. Surface-pump hydraulic dredge, limited to 200 feet.
3. Wire line dredges (grab bucket, clamshell, orange peel, etc.), limited to 500 feet.
4. Air-lift dredge.

Although these dredge types have functioned successfully at much greater depths, the depth limits given above are those for the majority of present day operations.

For deep-sea areas, the deep-sea drag dredge and the deep-sea hydraulic dredge have been envisioned, which might prove practical for mining phosphorite and manganese nodules. The cost of mining a ton of nodules by deep-sea drag dredge was estimated by Mero[27] to range from about $12 at a depth of 1,000 feet to more than $40 at 10,000, provided the nodule concentration is 1 pound per square foot. The use of deep-sea hydraulic dredging, on the other hand, lowers these figures to an acceptable range of $2.29 per ton at 1,000 feet and roughly $5 at 20,000 feet.

The most recent breakthrough in sea-floor dredging is a simple system of air-lift dredge (Figure 5) successfully tested in July 1970 on the Blake Plateau, approximately 170 miles off the coast of Georgia and Florida. From depths between 2,400 and 3,000 feet, the dredge succeeded in extracting a continuous flow of nodules from the ocean floor. The system on board

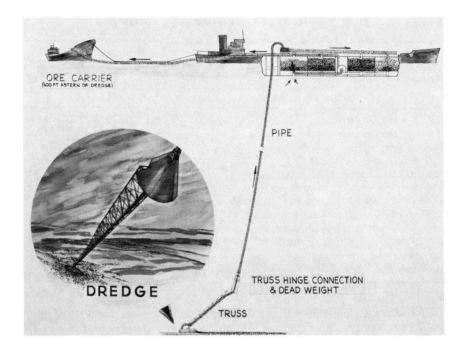

Figure 5. The air-lift dredging system employed by Deepsea Ventures, Inc., for scooping and lifting manganese nodules from deep ocean floor. *Adapted from illustrations by Deepsea Ventures, Inc.*

the *Deepsea Miner* was operated by Deepsea Ventures, Inc., a subsidiary of Tenneco, Inc. It represents approximately a decade of research and an investment of nearly $18 million. The objective following this initial success is the capability to operate at greater depths between 18,000 and 20,000 feet, probably in the Pacific Ocean where nodules are plentiful. With a processing plant planned for the mid-seventies, the operators hope to attract an investment of $150 to $200 million to finance full-scale production before the end of this decade. The Japanese ship *Chiyoda Maru No. 2* has succeeded in recovering nodules from comparable depths in the vicinity of Tahiti by using a continuous line bucket system.

While this manuscript was being readied for the press, a breakthrough was achieved in processing manganese nodules. After investigating more than 100 processes, Deepsea Ventures, Inc., reported in the June 1971 issue of *Ocean Industry* magazine that it has developed a chemical hydro-metallurgical process to extract economically manganese, copper, cobalt, and nickel from seabed nodules (Figure 6). The expected yield from full-scale processing will be 260,000 tons of manganese; 12,600 tons of nickel; 10,000 tons of copper; and 2,400 tons of cobalt.

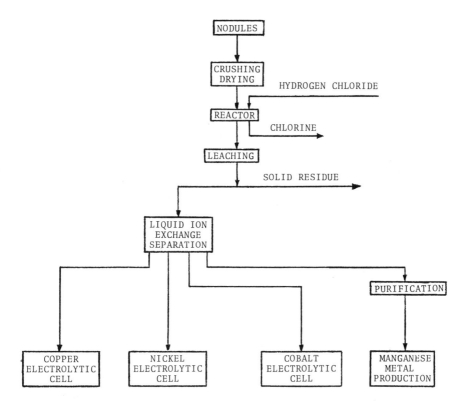

Figure 6. Flow diagram illustrating the hydrochlorination process of Deepsea Ventures, Inc., for extracting the copper, nickel, cobalt, and manganese components of manganese nodules dredged from the ocean floor. The process was successfully demonstrated in a pilot-plant operation. *Courtesy of Deepsea Ventures, Inc.*

Such technological developments point up the fact that technology is a product of incentive and time. It is within present knowledge to acquire the desired technological capability to exploit the seabed more fully, but the time factor and the speed of development are dependent on the incentive to invest the required capital. This incentive is a combination of economic, social, and political factors.

Phosphorite

Phosphate rock on the seabed, to justify the operational costs of its recovery, has to compete with present land deposits in purity and grade of the ore body, abundance, cost of transportation, processing and beneficiation, and in cost of exploration at sea versus on land.

Although most of the phosphate deposits on land are of marine origin, their exposure to weathering processes through geologic time results in an enriched final product with a higher content of phosphate than the marine deposits. Submarine pellets and nodules seldom contain more than 30 percent P_2O_5 whereas the cutoff grade for fertilizer production is about 31.5 percent, and land deposits now being worked contain around 35 percent.

World demand for phosphate products (industry 20 percent, fertilizer 80 percent) has been increasing at an annual rate of 6 percent. In 1965 world consumption totaled 63 million metric tons; in 1975 a minimum of 120 million tons will be needed, and in the next decade the world demand is expected to double.[28] Earlier calculations which show the expected increase in world population, and the per capita consumption projected to the year 2000 A.D., are shown in Tables 6–9.[29] As indicated in Table 8, by the year 2000, the total world consumption of phosphate rock is forecast to reach 7.6 billion long tons, containing about 2.28 billion long tons of phosphate, and total U.S. consumption is forecast to reach 1.23 billion long tons, containing about 380 million long tons of phosphate. The world reserve and potential resources of phosphate rock and apatite were reported by the U.S. Bureau of Mines to contain nearly 50 billion long tons of phosphate.[30] This supply is believed more than adequate for projected demands. Furthermore, new land deposits are being discovered,* and new processing and more efficient beneficiation are being applied in developing known reserves.

Transportation costs play a major role in the economics of phosphates. Shipping phosphate rock from Florida to California, for example, costs $4.50 a ton; rail transportation from Idaho is much higher. With transportation costs added to production costs, the price of Florida phosphate rock in California approaches $12 a ton, and in Japan about $14 per ton. For this reason, submarine phosphorite deposits in certain areas like the Thirty- and Forty-Mile Banks off San Diego and the phosphorite sand deposit in Santo Domingo Bay, Baja California, even if mined at costs higher than those on land, can have promising and important potential advantage. These deposits can compete with Florida land deposits for Mexican, Japanese, Australian and Pacific markets, if they can be mined at recovery and primary beneficiation costs less than $8.50 per ton.[31] The price of phosphate rock at Florida mines is about $7 per ton. Aside from this advantage for special localities, transportation cost is considered a disadvantage for

---

* A new discovery in Saudi Arabia was made by geologists of the U.S. Geological Survey near the Jordanian border. The Turayf Area I covers 1,300 square miles and is estimated to hold about 1.1 billion tons. Thaniyat Turayf Phosphate Area II covers 1,500 square miles and contains almost 100 million tons of high-grade ore. (Charles R. Meissner, Jr., U.S. Geological Survey, "Phosphate Deposits in Sirhan-Turayf Basin," in *Mineral Resources Research 1967–8*, Directorate General of Mineral Resources, Saudi Arabia, pp. 52–53.

TABLE 6.—Population Projections, 1965–2000

	Population in Millions			
United States, South Asia, Africa and Latin America, World	*Mid-1965 Estimated Population*	*1970 Continued Trends*	*1980 Continued Trends*	*2000 Continued Trends*
World	3,308	3,626	4,487	7,410
United States	194	211	252	362
Percent	(5.8)	(5.8)	(5.6)	(4.9)
Africa, South Asia, and Latin America	1,534	1,724	2,263	4,214
Percent	(46.4)	(47.5)	(50.5)	(56.8)
Remainder of world	1,580	1,691	1,972	2,834
Percent	(47.8)	(46.7)	(43.9)	(38.3)

Note: Percentage in ( ) is percentage of total world population.

Source: *Population Bulletin*, Vol. XXI, No. 4, October 1965, by Population Reference Bureau, Inc. Although the U.S. population figures and projections are weakened somewhat by more recent census data, the overall trend remains essentially valid.

TABLE 7.—Per Capita Consumption (Apparent) of Phosphate Rock

	United States		World	
Year	*Apparent Consumption[1] (Long Tons)*	*Pounds Phosphate Rock per Capita*	*Production[2] (Long Tons)*	*Pounds Phosphate Rock per Capita*
1960	13,337,000	166.0	41,170,000	31.0
1961	14,058,000	——	44,780,000	——
1962	15,260,000	——	47,320,000	——
1963	15,474,000	——	50,590,000	——
1964	16,546,000	192.0	57,910,000	40.0

[1] Apparent consumption is measured by phosphate rock sold or used plus imports minus exports.
[2] Production is used as no consumption figures were available in the sources of information. It is assumed that world production and consumption are nearly equal.

Source of consumption and production figures: *Minerals Yearbook*, Vol. I (1964 edition).

worldwide marketing of submarine phosphorite. On the other hand, logistics are sometimes outweighed by efficient mining practices and particularly marketing. A case in point is that Florida producers are able to sell large tonnages of phosphate rock to West Germany and Italy despite the large nearby reserves of North Africa, particularly Morocco (21 billion tons of P_2O_5).[32]

The Hashemite Kingdom of Jordan possesses an excellent grade of phosphate, but the industry has suffered from logistics and marketing problems. Political factors also plagued the Jordanian industry in its China

TABLE 8.—Projected Total Consumption of Phosphate Rock, 1966–2000
[In tons of 2,240 pounds]

	Total Long Tons Consumed Based on	
	Continued Trends in Population[1].	*Medium Trends in Population*[1].
United States:		
192 pounds per capita ..	804,500,000	770,000,000
	$(249,000,000\ P_2O_5)$	$(238,000,000\ P_2O_5)$
300 pounds per capita ..	1,225,000,000	1,220,000,000
	$(380,000,000\ P_2O_5)$	$(378,000,000\ P_2O_5)$
World:		
40 pounds per capita ...	3,225,000,000	2,895,000,000
	$(967,000,000\ P_2O_5)$	$(869,000,000\ P_2O_5)$
100 pounds per capita ..	7,600,000,000	7,100,000,000
	$(2,280,000,000\ P_2O_5)$	$(2,130,000,000\ P_2O_5)$

[1] Based on population trends shown on table 6.

TABLE 9.—Reserves and Potential Resources of Phosphate Rock in the
United States
[In millions of long tons]

	Total Reserves		Potential Resources	
Source	*Marketable Product*	P_2O_5 *Content*	*Marketable Product*	P_2O_5 *Content*
Arkansas	——	—	20	5
Florida	2,040	660	23,350	4,932
North Carolina	[1]2,000	[1]600	([2])	([2])
South Carolina	——	—	9	2
Tennessee	80	12	5,398	1,129
Western field[3]	3,000	870	20,000	5,800
Total (rounded) ...	7,100	2,100	49,000	12,000

Note: A large tonnage of lower grade phosphatic material in the western field is not included in resources. The Florida and Tennessee potential resources include the low-grade minable material.

[1] Estimate.
[2] Data not available.
[3] Includes Idaho, Montana, Utah, and Wyoming.

Source: *Mineral Facts and Problems* (1965 edition), p. 704.

market. When Jordan voted with the West against seating the People's Republic of China at the United Nations, China retaliated immediately by closing its market for Jordan's phosphate. Similarly, when Jordan's borders with its neighboring Arab countries were closed in 1971, following clashes with the Palestinian guerillas, exports to Turkey and Yugoslavia came to a halt. There were indications that Turkey and Yugoslavia were ready to accept Israeli phosphates instead. In order to ward off that eventuality, Jordan rushed to pay the difference in the cost of shipping Moroccan and Tunisian phosphates to Turkey and Yugoslavia to insure the continuous flow and thereby safeguard its interests. The cost of this operation to the Jordanian government was $90,000.[33]

The most serious deterrent to offshore phosphorite mining is a lack of knowledge of the marine environment. The lack of experience and a preference for land exploration and exploitation deter the decision makers. This is particularly true of small entrepreneurs whose financial resources are too limited to permit risk-taking. Nevertheless, regardless of the arguments and conflicting opinions, the submarine phosphorite deposits seem to have potential economic value, the economic exploitation of which is only a matter of time.

Manganese Nodules

Manganese is used as an additive metal in the manufacture of steel to reduce its brittleness. It seldom costs more than 5 cents per pound and is, therefore, much cheaper than other additives that can perform its functions. Steelmaking accounts for more than 95 percent of manganese consumption. Although the United States is the largest world consumer, world consumption is expected to rise with the development of steel plants in the emerging nations. The United States consumes approximately one-seventh of the world's ore, importing 99 percent of its needs (35 percent or more manganese content), with domestic stocks enough for about six months. Domestic supplies are inadequate, and low-grade recoverable ores are too expensive to process. Using the cheapest of the tested processes, and allowing for further technological progress, it would cost the United States about $1 extra per ton of steel to turn either to domestic deposits or to slags for manganese.[34]

The extent of world manganese reserves is not known; however, they are so large and high grade that at the present rate of consumption they can be considered virtually unlimited. This is the reason for the assertion that submarine manganese nodules can be mined profitably for their minor constituents, particularly copper, nickel, and cobalt.

The average high-grade nodules contain 35 percent manganese, 2.3 percent copper, 1.9 percent nickel, and 0.2 percent cobalt (see Table 10). Brooks used the percentages 35, 2, 2, and 0.5, respectively.[35] One ton of such nodules, he estimated, contains 700 pounds of manganese, 10 pounds of cobalt, 40 pounds of nickel, and 40 pounds of copper. The average 1970 market prices of these components were:[36] manganese, 2.5 cents per pound of ore (33 cents per pound of metal); cobalt, $2.20; nickel, $1.30;

TABLE 10.—Reserves of Metals in Manganese Nodules of the Pacific Ocean

Element	Amount of Element in Nodules (Billions of Tons)[1]	Reserves in Nodules at Consumption Rate of 1960 (Years)[2]	Approximate World Land Reserves of Element (Years)[3]	Ratio of Reserves in Nodules/ Reserves on Land	U.S. Rate of Consumption of Element in 1960 (Millions of Tons per Year)[4]	Rate of Accumulation of Element in Nodules (Millions of Tons per Year)	Ratio of Rate of Accumulation Rate/ of U.S. Consumption	Ratio of World Consumption/ U.S. Consumption
Magnesium	25.0	600,000	[5]L	—	0.04	0.18	4.5	2.5
Aluminum	43.0	20,000	100	200	2.0	.30	.15	2.0
Titanium	9.9	2,000,000	L	—	.30	.069	.23	4.0
Vanadium	.8	400,000	L	—	.002	.0056	2.8	4.0
Manganese	358.0	400,000	100	4,000	.8	2.5	3.0	8.0
Iron	207.0	2,000	[6]500	4	100.0	1.4	.01	2.5
Cobalt	5.2	200,000	40	5,000	.008	.036	4.5	2.0
Nickel	14.7	150,000	100	1,500	.11	.102	1.0	3.0
Copper	7.9	6,000	40	150	1.2	.055	.05	4.0
Zinc	.7	1,000	100	10	.9	.0048	.005	3.5
Gallium	.015	150,000	—	—	.0001	.0001	1.0	—
Zircon	.93	100,000	100	1,000	.0013	.0065	5.0	—
Molybdenum	.77	30,000	500	60	.025	.0054	.2	2.0
Silver	.001	100	100	1	.006	.00003	.005	—
Lead	1.3	1,000	40	50	1.0	.009	.0009	2.5

[1] All tonnages in metric units.
[2] Amount available in the nodules divided by the consumption rate.
[3] Calculated as the element in metric tons. (U.S. Bureau of Mines, Staff, 1956.)
[4] Calculated as the element in metric tons.
[5] Present reserves so large as to be essentially unlimited at present rates of consumption.
[6] Including deposits of iron that are at present considered marginal.

Source: From Mero, *op. cit.*, p. 278.

and copper, 60 cents per pound. Thus the gross value of one ton of manganese nodules is $115 (at 1963 prices, Brooks estimated $83 per ton, with a range between $45 and $100, depending on variation in composition).

The amount to be mined will have an immediate effect on the current prices. Mero[37] estimated that if an operation were designed to mine an average grade of the nodules to produce 100 percent of the U.S. consumption of nickel, that operation would also produce about 300 percent of its annual consumption of manganese, 200 percent of that of cobalt, 100 percent of that of titanium, 300 percent of that of vanadium, and about 500 percent of that of zirconium. This sort of calculation has been repeated by Francis L. LaQue, who assumed a nodule composition of 30 percent manganese, 1 percent nickel, 0.75 percent copper, and 0.25 percent cobalt.

If 100 percent of the world's copper needs were filled from such a source, there would also be produced 133 times the world consumption of cobalt and 15 times the world consumption of nickel.[38]

These calculations, however, did not take into consideration the gross returns and the effects of massive mining on current prices. Brooks, using his 35/2/2/.5 percentages again,[39] estimated that 2,000 to 5,000 tons of nodules mined per day would drop manganese prices from 90 cents to 50 cents per unit. At this price several of the large African and South American producers could continue to operate, but Indian mines and most smaller producers would probably be forced out of the market. The ton of rich ore that was valued at $83 would drop below $64. There would follow an annual decrease in profits from $26 million to $14 million (production rate 2,000 tons per day), and from $65 million to $36 (at 5,000 tons per day).

Similar warnings about the adverse effects of nodule mining on world prices were echoed recently in a study of the economics of manganese nodules. The study concluded that realistic production rates will severely affect the world market price of cobalt; will have a significant effect on manganese and nickel prices; and will cause little or no changes in the world copper market. The study made projections into the years 1985 and 2000, and concluded further that "the exploitation of manganese nodules could be detrimental to the economies of specific developing countries with manganese, nickel or cobalt mining industries." *

Regardless of the arguments for or against mining submarine manganese nodules, there is general agreement that these nodules are a tremendous potential resource. Brooks concluded that:

> Though the claims for the returns from deep-sea mining of manganese nodules have been exaggerated, this by no means eliminates them as possible manganese sources. To the contrary, it is my conclusion that they are the only alternative source that is likely to be developed in the middle-term future.[40]

COMMERCIAL RECOVERY OF OFFSHORE PETROLEUM

The origin and occurrence of petroleum and the reservoir parameters requisite to its accumulation were described in an earlier section. In the pursuit of oil and gas, the first step is to search for it. Geological exploration is followed by exploratory drilling to see if the structures discovered actually contain commercially producible oil. If they do, the next stage is the development or exploitation.

* Bollow, George Edward. *Economic effects of deep ocean minerals exploitation.* M.S. thesis, U.S. Naval Postgraduate School, Monterey, California, Sept. 1971, p. 87.

Exploration

In the search for oil, exploration techniques employ a variety of sophisticated equipment necessary for geophysical investigations. Modern echo-sounding and seismic equipment are capable of recording the profile of the sea floor and the shapes of the underlying sedimentary strata. These devices vary with the type of work to be performed and the desired resolution, accuracy, depth, and areal extent. Data are interpreted with the aid of computers, and computerized operations are becoming routine.

The geophysical survey is of no practical use if the structure discovered under water cannot be returned to, located, and occupied for drilling and development. The location of potential and actual drilling sites is becoming progressively more dependent on geodetic positioning techniques as the oil industry moves farther offshore. The fluid nature of the surface element— the superjacent water—compounds this difficulty. To solve this problem, several methods have been developed to use satellites directly as control for surveys at sea.*

Drilling

A major element in the exploration for petroleum is the exploratory drilling that follows the geophysical work. This drilling determines the economic value of the discovery. A structure that has all the elements requisite for the accumulation of hydrocarbons may, when drilled, produce a dry hole; hence the old saying among oilmen that "oil is where you find it."

Offshore drilling equipment is a direct descendent of equipment for land drilling, with adaptation to the marine environment. The basic rig

---

* Alton B. Moody, "Geodesy and Oil Exploration," in *Papers from the Technical Conference of the American Society of Photogrammetry—American Congress on Surveying and Mapping, October 7–10, 1970,* Denver, Colorado, pp. 301–312, Moody states that as operations extended seaward beyond the range of visual observations, shore-based radio systems were pressed into service, such as shoran, Raydist, lorac, and Decca, which are still widely used. These, however, suffer from a number of limitations, including propagation problems, geometry, limited range, logistics, and political difficulties. Errors vary from 500 meters to 100 meters (roughly 1,700 to 330 feet) in locating a position. The most sophisticated system thus far developed is now in operational use by the Western Geophysical Company of America. It consists of a satellite receiver, Doppler sonar equipment, an inertial navigator, and various ancillary equipment. Post analysis of data permits determination of the position of the vessel during the survey to an accuracy of about 150 feet on the continental shelf, and with an error about 10 percent larger in deeper water. Thus, the system exceeds the accuracy requirements specified for oil exploration at sea.

(See also: Alton B. Moody and W. A. Knox, "Geodetic Position-Finding at Sea and the Search for Oil," *Surveying and Mapping,* December 1970, pp. 581–591.)

(including derrick, kelley, rotary table, blowout preventer, drill pipe, bit, and casing) is used from several types of platforms. There are fixed platforms constructed on piles driven securely into the ocean floor. Most fixed platforms have a maximum efficiency in waters less than 300 feet deep, with more recent designs reaching 600 feet. A second stage of evolution is the semifixed platform designed to rest on the sea floor while drilling takes place; when drilling is finished, the platform is refloated and moved to another site. The design that employs the jack-up concept is the most versatile and most common among offshore platforms. The third type is the floating platform. Floating platforms are far less costly for exploratory drilling than the fixed or semifixed platforms. They also have the added advantage of capability in much deeper water. The basic concept is that of a conventional rig in the center of a ship or barge modified to perform the task. Recent drill vessels can operate without being anchored, by dynamic positioning over the drill site.

Although exploratory drilling has been conducted mostly in depths less than 1,000 feet, and completion of producing wells in less than 300 feet, further advances in both technologies are imminent. A wildcat for a commercial well was recently drilled from a drill barge off Santa Barbara, California, for Humble Oil Company (Tract 322) in 1,500 feet of water, in the vicinity of a prolific strike (Humble/Stancal Tract 325) in 1,050 feet of water. Humble Oil has also announced plans to install a 60-well platform for drilling and production in 700 feet of water in the Santa Barbara Channel. With exploratory drilling it is easier to achieve added depth capability than with production drilling, because once the hole is drilled no more equipment is needed as in the production wells. Reentry is a basic requirement in drilling for petroleum. At certain intervals, when the bit wears out, a new bit has to replace the old bit. In order to accomplish this, the whole pipe length in the hole has to be withdrawn, the old bit is replaced, and the drill string is reintroduced into the hole. Guidance back into the hole is hard to achieve in deep waters.

In research drilling, no reentry into the hole is required. This is why the *Glomar Challenger* was able to drill and core in 11,720 feet of water on Sigsbee Knolls, and later set a record in the North Atlantic, drilling 2,759 feet below the ocean floor in 16,316 feet of water. By the end of 1971, the *Glomar Challenger* broke that record, too, drilling in more than 20,320 feet of water to 1,237 feet, about 800 miles southeast of Tokyo. Powering about 4 miles of drill pipe from a floating vessel is no small feat! However, as evidence of the rapid development in drilling technology, a breakthrough was achieved by the *Glomar Challenger* when, on June 14, 1970, a deepwater hole was successfully reentered. The crew changed the worn-out bit and succeeded in finding and reentering the hole in 10,000 feet of water. Several months later, a similar operation was successfully completed in

13,000 feet of water; the bit wore out after drilling 2,300 feet into the sea floor, was changed and reentered into the same hole to drill 200 additional feet. Reentry was accomplished with the aid of a high-resolution scanning sonar system looking through the drill bit and guiding it into a funnel-shaped receiving cone mounted on the ocean floor, and a system for steering the drill pipe toward the cone (Figure 7). This achievement heralded a new era in offshore technology, but only for drilling, not for production. The aim, however, is production—a challenge more formidable, though not insurmountable.

Production

Following successful and promising exploratory drilling, a well is completed and equipped for production. Production requires the installation of a well head, a valve complex often called the "christmas tree," flow lines to move the oil to the separators, separators to separate the gas from the oil, and pipelines to transfer the products to storage tanks and refineries.

Some of this equipment is installed on platforms above water at shallow depths to about 300 feet, and on the ocean floor in deeper waters. There are problems of installation, production, and servicing. The last is a major activity which is performed periodically throughout the life of the well. Oil-well ancillary services involve a number of complicated activities often critically limited by water depth. One such activity includes the services of divers, underwater submersibles, and the attending support vessels and equipment.

The majority of above-water production facilities is effective for wells in waters no deeper than 340 feet, although new designs have pushed this limit to twice the depth. As water depth increases, however, drilling costs increase drastically (see Appendix 10). In the Gulf of Mexico, for example, the cost for existing platforms rises from $1.5 million in 100 feet of water to $4 million in 350 feet, and an expected $12 million in 600 feet.*

The cost of drilling and completing a platform well rises from $200,000 in 100 feet of water to $425,000 in 350 feet. Within 1971, costs of drilling rigs and other equipment jumped 25–30 percent. Adding a share

---

* John L. Kennedy, "Offshore-Rig Construction Costs Will Continue to Climb," *Oil and Gas Journal,* March 16, 1970, pp. 136–140. Recent Federal changes in offshore regulations (expanded OCS Order No. 8) require safety and antipollution equipment which is believed by the oil industry to increase production costs about $150 million. Annual pollution-control expenditures by petroleum companies in the United States totaled $271.4 million for 1966, $357.9 million for 1967, and $381.6 for 1968. (*An Interim Report on Current Key Issues Relating to Environmental Conservation— The Oil and Gas Industries,* prepared by the National Petroleum Council's Committee on Environmental Conservation—The Oil and Gas Industries, June 22, 1970, p. 5.)

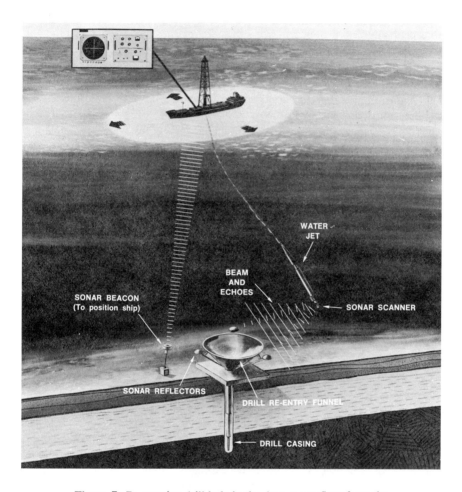

Figure 7. Re-entering drill hole in the deep ocean floor from the *Glomar Challenger.*

of the platform cost, the cost of drilling and completing a 12,000-foot under-water exploratory well is $550,000 in 100 feet of water. When projected to 600 and 1,000 feet of water the cost rises to $990,000 and $1,100,000 respectively (see graphs in Appendix 10). With the exception of the Santa Barbara Channel, areas deeper than 600 feet will present formid-able problems for production, and the size of the field will determine the economics of production. In the Gulf of Mexico, for example, a typical field is 200–300 million barrels. In depths of 1,200–1,300 feet, as in the Santa Barbara Channel, a field in the Gulf of Mexico will have to be close to a billion barrels in order to economically offset the costs in equipment, bad weather, and other hazards.

As petroleum operations are conducted in progressively deeper water, they require more and more sophisticated equipment. Remote control devices are an emerging necessity. Such refinements add significantly to the cost of recovery. One very important factor in underwater operations is the mobility and overall efficiency of divers. The deepest working dive on re-cord was to 700 feet, and laboratory simulated dives have exceeded 1,700 feet. In April 1970, divers from the British Royal Navy succeeded in mak-ing a "dry dive" to 1,500 feet. Later in 1970, two French divers participat-ing in project Physalie 5 made a record dive to more than 1,700 feet. The project is aiming at depths close to 2,000 feet.

Saturation diving capabilities have been extended at a rate of approxi-mately 125 feet per year during the past six or seven years (see Figure 12 in Appendix 11), but diver capability will probably be limited for the fore-seeable future to less than 1,500 feet. Furthermore, new drilling and com-pletion systems have been developed to minimize the need for divers.

Future Trends

It is hard to keep up with the speed at which the offshore oil technology is advancing; what is postulated as an artist's conception today may become a working model by the time these words appear in print. As water depth increases, it becomes necessary to abandon above-water plat-form equipment and resort to bottom installation and production systems.

Underwater wellheads to control the flow of oil or gas from a well have been installed in numerous locations around the world, some connected directly to shore facilities. An interesting concept in producing oil and gas in deep water is represented by WODECO's underwater sphere for drilling and production.* This sphere includes a shirt-sleeve environment at a sub-

---

* Western Offshore Drilling and Exploration Co. (Fluor Drilling Services, Inc.). The company's project manager indicated that the concept is still under development, and has undergone changes in design. Its use depends on favorable conditions other than technology.

merged depth of 150 feet (25-pound pressure per square inch), wellheads, separating, metering, and pumping equipment. It is designed to be effective in water depths exceeding the 1,300 feet which is the oil industry's immediate target for operating depths. This target depth is directly related to known resources in the California offshore fields, which illustrates the influence of discovery on the push for technological development.

A similar spherical habitat has been tested as part of a prototype seabed oil-production system designed by Lockheed Offshore Petroleum Systems (Figure 8). The manned capsule (Figure 9) is designed for a one-atmosphere environment at depths of 1,200 feet, and the whole system can be extended to 2,000 feet. The system includes wellhead cellars, pipeline assembly, manifold center on the sea floor, and gathering lines that carry the petroleum either to subsea storage or directly to surface separators.

Separators are systems that separate the gas from the oil to facilitate pumping the oil to storage facilities ashore or to mooring tankers. The most recent application of a subsea production system was undertaken by the Dubai Petroleum Company in the Arabian Gulf (Figure 10). The need for this subsea production system arose following the discovery in June 1966 of the Fateh Field about 60 miles off the shore of Dubai. To develop this field,

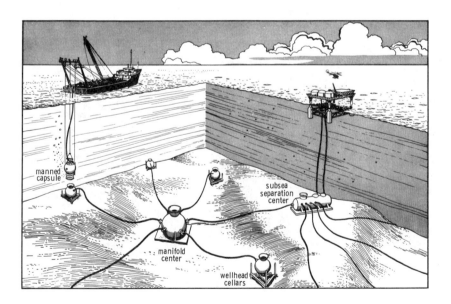

Figure 8. Conceptual design of an underwater petroleum production system. The system involves construction and emplacement on the ocean floor of man-rated pressure hulls containing normal oil-field components. Components are serviced by manned capsule (see Figure 9). *Adapted from material supplied by Lockheed Petroleum Services, Ltd., British Columbia, Canada.*

Figure 9. Service capsule (Top sphere) being lowered to couple with wellhead cellar (Bottom sphere) encapsulating production equipment. The service capsule provides transportation from the surface to the ocean floor, and contains life-support and oil-field equipment. The one-atmosphere, shirt-sleeve environment eliminates the need for highly specialized diving personnel and support facilities. *Photo courtesy of Lockheed Petroleum Services, Ltd., British Colombia, Canada.*

Figure 10. Artist's impression of offshore oil-field development showing underwater production, separation, and storage systems. *Photo courtesy of British Petroleum Company, Ltd.* (1969).

the company would have had to lay pipeline all the way to shore for storage then lay more pipelines back from shore storage facilities to reach water 10 miles offshore deep enough to accommodate tankers for loading. The whole production system was exhibited at the Oceanology International 1969 exhibit. The storage tank was emplaced in August of that year. It was built at a cost of about $7 million by the Chicago Bridge and Iron Company. In April 1969, British Petroleum Company Ltd. [personal communication] was planning to test a limited subsea production scheme involving an oil and gas separation unit on the seabed elsewhere in the Arabian Gulf. This separator was emplaced in August 1970 on well No. 32, with all processing equipment resting on the sea floor. The separator started processing crude oil initially at 5,000 barrels per day, separating oil and gas, measuring them, and discharging the oil into a pipeline to shore. This development illustrates further the speed with which petroleum technology is advancing from the conceptual stage to the operational stage. In 1969 the storage tank, Khazzan Dubai No. 1, was placed on the sea floor 58 miles from shore, and in 1970 the first self-setting oil and gas separator was successfully emplaced on the sea floor in the Zakum Field near Das Island, under 75 feet of water.

These are only a few of the many approaches to the development of technology for the recovery of offshore petroleum. To minimize reliance on divers, added emphasis is being given to the design and use of submersibles. Submersibles have been used in underwater oil-field operations, and offer a potential to perform at virtually unlimited depths. It is estimated that technological capabilities for production will be achieved within five years to the same depths of 1,300 or 1,500 feet already within the capabilities of exploration drilling. Even before the middle of this decade it is expected that the *Glomar Challenger* will be able to obtain a seabed core 5,000 feet long in 30,000 feet of water.

Beyond the mid-seventies and toward the end of this decade, a total of 500 offshore drilling rigs are expected to be in operation around the world. Technology will have reached the point at which the water depth is no longer the determining factor. Factors of economic and political feasibility will then play the decisive role in formulating policy for offshore exploration and exploitation.

Supply and Demand

The "free world" consumption of liquid petroleum for 1969 totaled 37,192,000 barrels per day (b/d), representing an increase in demand by 8.4 percent over the previous year.[41] During the first nine months of 1970 it exceeded 40 million barrels per day (mb/d), a growth of more than 9 percent. For the United States the demand for oil recorded a 5 percent increase; for the rest of the "free world" the demand growth was almost 12 percent.[42] Projections for the 1980s indicate that, at an average rate of growth of 7 percent annually, world consumption of oil will be nearly four times that of today, and the use of petroleum gas will increase as much as five times in the same period.* (For estimated U.S. domestic demand for liquid hydrocarbons, see Appendix 12.)

Against this demand, world production is variously estimated at between 34.4 mb/d and 41.3 mb/d.† Offshore production accounts for 16–18

---

* Lewis G. Weeks, "The Gas, Oil and Sulfur Potentials of the Sea," *Ocean Industry,* June 1968, p. 43. Of this world consumption, the Communist countries take 16–17 percent. In the first quarter of 1971 (as reported in the *Oil and Gas Journal,* May 31, 1971, p. 18), the Communist production averaged 7.874 mb/d.

† Weeks, *op. cit.,* put current (1968) production at 35.3 mb/d; The *Oil and Gas Journal* (December 29, 1969, p. 95) estimated the 1969 production at 41,266,100 b/d; and The Chase Manhattan Bank (Sparling *et al., op. cit.,* p. 95) gave the figure of 34,390,000 b/d for crude oil production for the "free world." J. D. Moody gives a "best guess" for 1990 as 98 mb/d. ("Petroleum Demands of Future Decades," *American Association of Petroleum Geologists Bulletin,* December 1970, pp. 2239–2249.) Figures as of June 1970. More recently (June 14, 1971), Weeks was reported to have revised his estimates to a world production of 43 mb/d of which the offshore production represented 17 percent (*Ocean Oil Weekly Report*), while Larry Auldridge (*Oil and Gas Journal,* May 31, 1971) forecast world production to average 50 mb/d for 1971.

percent of the world's total. Weeks has predicted that by 1978 an offshore yield of 23 mb/d is expected, representing 33 percent of a world total of 70 mb/d. He also estimated that proved world oil reserves total 425 billion barrels, which would last at least through this century. Moreover, the ultimate world potential of all resources *offshore* totals 1,600 billion barrels. For comparison, ultimate world potential of comparable resources *on land* was estimated at 4,000 billion barrels.

Despite this enormous land potential, the oil industry is vigorously delving into the offshore fields. Current investments in offshore operations were reported by the Department of the Interior to be nearly $20 billion. Investment is expected to increase at the rate of $3 billion annually, reaching a total of $50 billion by 1980. Estimates given by L. G. Weeks show a "probable total upward of $25 billion" up to 1968, expected to reach $50 billion by 1978.

Operations on land are generally less costly than those offshore. More costs are added to offshore operations as the new consciousness of environmental concern gains momentum. Hazards of offshore operations include those encountered in land operations, aggravated further by the marine environment, plus a new breed of hazards peculiar to the underwater world. Safety and antipollution requirements have already added a heavy burden to the industry's outlays, and more stringent regulations will add further to the spiraling costs of penetration into deeper water.

Although exploration expenditures offshore are less than those for land, drilling and production make up for this margin. Technological innovations often tend to be glamorous, and their novelty tends to overshadow and supersede older and more reliable technology. In the words of Eduardo J. Guzman:

There are many examples in the world of these premature adventurous offshore campaigns involving the use of costly geophysical methods where less expensive exploration approaches still could yield considerable success in the discovery of new reserves. This worldwide trend is not new in the history of exploration. It has happened repeatedly even within the United States, where every new method or tool has tended to displace all other previous ones, and usually at higher operating costs. . . . Marine or offshore exploration, particularly involving seismic work, is easier, faster, and cheaper than almost any method on land, but we tend to forget that offshore drilling and production are several times more expensive." [43]

Besides exploration, one is further faced with the following factors:
—Skyrocketing costs of lease sales and bonuses, particularly in the United States ($600 million each in Santa Barbara Channel and Gulf of Mexico in 1968, and the December 1970 sale of $850 million in the Gulf of Mexico);
—Higher overall costs for offshore operations;

—Projected increase in future expenditures toward deeper water;

—Higher costs eventually passed on to the consumer; and

—Offshore hazards like blowouts, fires, and oil spills increasing in frequency and magnitude.

The question is: Has the industry exhausted land resources? The answer, of course, is no. The land potential of 4,000 billion barrels of oil estimated by Weeks does not include the vast amounts of synthetic petroleum in bituminous rocks such as oil shales and tar sands. The U.S. Bureau of Mines (Information Circular 8425) estimates that the Green River Formation oil shales contain 2 trillion barrels of oil: 800 billion barrels at 15 gallons of oil for a ton of shale, and about 500 billion barrels at 25 gallons of oil per ton of shale are considered practically exploitable. The proved recoverable reserves alone are about four times the total proved U.S. petroleum reserves.* In the United States, about 80 billion barrels of oil from the more accessible high-grade deposits of the central Rocky Mountains can be considered available with demonstrated methods of extraction, and at costs approaching the present-day costs of petroleum of comparable quality. The development of these deposits, as Weeks put it, has been caught in "the political jungle that has invaded the outlook, and which is partly responsible for the lack of progress in this field." He also adds: "Perhaps the petroleum industry on its part has not supplied the kind of energetic and particularly united leadership required in dealing with government." [44]

Eventually, however, technological development in the extraction of oil from shale will make it competitive with conventional oil and gas, at prices estimated at between $3.75 and $5.25 per barrel of oil. With the enormous deposits of oil shale available, such prices are certain to halt the spiraling costs of producing oil from conventional sources. This ceiling becomes a serious deterrent to any advance into deeper waters offshore, as well as costly production from sources on land.

Other than oil shale, there are still greater amounts of potential synthetic oil and gas in coal. It is estimated that the world has enough coal to last 1,500 years. Then there are tar sands, nuclear power supplemented in time by the virtually inexhaustible fast-breeder reactor, and other land sources of energy (possibly even fusion) which will eventually be developed and become competitive with oil and gas. There are also those regions on land that have not been explored, and new discoveries like the Alaska

---

* Several countries are already exploiting their oil-shale deposits. At the 1970 International Gas Conference in Moscow, the U.S.S.R. revealed that improved mining and mechanized handling methods have pushed Soviet oil shale beyond 22 million tons per year. (*Oil and Gas International*, September 1970, page 117.) Australia also has announced a new project to begin production in 1974 from its vast oil-shale reserves; the reserves indicate a field life of 50 years.

North Slope are not to be ruled out. Some scientists believe that a very high percentage of the land surface is still unexplored and is potentially promising.

The oil industry contends that despite higher initial investment, the development of an offshore field eventually reduces offshore costs. Offshore platforms can drill 50–60 wells from one location; oil accumulations in younger strata offshore provide greater yields and higher success ratios. When distributed over the entire operation, these costs are expected eventually to be lowered to reasonable and acceptable levels.

5

Policy for Seabed Resources

Given the abundant resources of the seabed and the challenges they present, especially to the dynamic petroleum industry, what are the implications of this situation for national policy and international diplomacy? Before the complex issues of global agreements for apportionment of the ocean's resources are discussed, some considerations of U.S. national interests and policy should be reviewed.

POLICY FOR SUBMARINE MINING

A foremost consideration, present in all mining operations, is the necessity for maintaining an approximate balance of supply against demand. A prime economic characteristic of all minerals, except those that are scarce and precious, is their price sensitivity. Although supply may be elastic, demand is not. The housewife does not buy an extra tank of oil, and the steel mill does not double its order for manganese merely because the supply is plentiful. This characteristic is of particular significance for projects to mine the seabed or drill into it for oil. The higher initial cost of such operations needs to be evaluated against the possibility of a fall in prices such as would make the entire operation uneconomical after heavy capital investments have been poured into it.

Moreover, the disruptive effect of severe price fluctuations would extend far beyond the extractive industry directly involved. On the other hand, if it turned out that submarine mining was a lower-cost operation, the impact would be no less severe on conventional mining operations. The problem of overproduction, even of one component, would be analogous to a free import policy without safeguarding the local industry. Prices would go down, mining centers on land would close down, and whole communities might have to migrate in quest of livelihood elsewhere. As submarine mining would immediately relate to coastal areas, populations from inland would seek the already overpopulated coastal areas. Land, housing, equipment, business, and all such related activities left behind would feel the impact.

The fact that seabed deposits are being researched and seriously considered is enough to influence business decisions and put a firm ceiling on the long-run price of manganese, cobalt and nickel. When deep-sea mining becomes a reality, future prices might well be lower than today's and certainly would not exceed production costs.

The many unknowns surrounding offshore mining are a source of uncertainty causing decision makers to prefer land operations. Research dollars are directed by generally conservative mining companies toward more familiar and less risky land applications. Several mining economists believe that the high investment in the research and development of deep-sea mining will exclude most mining companies. They maintain that such development will come from larger, nonmining companies such as the oil and aerospace industries, or from consortia of several small companies. This is also the thinking behind the future production envisioned by Deep-sea Ventures, Inc., following its successful mining operation in 1970. The risk, the high capital investment, the unknowns, and the lack of experience in the marine environment will deter the small entrepreneur from venturing into the deep-sea operations.

Another deterrent stems from a source totally nontechnical and non-economic: that is the legal uncertainties. Most essential and fundamental to efficient mining practice is the exclusive right of the discoverer to exploit the minerals discovered and the security of tenure while he does so. The deep-sea operations of the future are in offshore areas undefined and undecided in proprietary and jurisdictional terms. Who owns the ore body that has been found far out in the middle of the ocean? Many companies have suffered from legal problems on land, and the failure of the world community to agree to an ocean regime is a serious deterrent to ocean mining ventures.

In the final analysis, it is the consensus that deep-sea mineral deposits are substantial and that they form a great potential resource. Once a substantially rich deposit is found, the technology to exploit it is likely to be developed promptly—much sooner than anticipated. A rich deposit and successful and profitable operation are enough to dissipate the doubts and overcome the deterrents now gripping the industry. The outlook is one of cautious optimism, but a legal regime must be established and international agreements effected before the mining industry ventures into the ocean deeps.

POLICY FOR OFFSHORE PETROLEUM

All considerations of present and future technological developments indicate that penetration into the deep-sea basins is only a matter of time. The capabilities that are not available today are certainly a short distance away. The question that is asked then is: Does technology justify expansion? In other words, just because the industry has the capability to do something, must it do it?

Certain inescapable and uncontestable facts cannot be ignored: (1) Petroleum is the basis of numerous activities, products, and needs throughout the world. (2) An increasing demand for petroleum products follows the increase in population trends toward urbanization, and the rapid industrialization of developing nations. (3) The numerous organizations engaged in petroleum activities and their ancillary services are organizations that, in the process of making profit, invest, generate, and circulate vast amounts of money which may shape the economies and politics of communities and even nations.

Regardless of the policies and arguments for or against the rapid expansion into the seabed, a few facts are clear: (1) Offshore operations are likely to continue at an ever-increasing pace. (2) Hazards are inherent in offshore operations, but can be avoided; otherwise, damage to the environment may be long-lasting or irreversible. (3) Present and future technology is capable of providing the needed petroleum and at the same time preserving the environment. (4) Numerous other enterprises besides the oil industry are engaged in activities using coastal waters.

On the international scene, many countries are plunging headlong into offshore exploitation with inadequate understanding of the disastrous hazards to their waters and shores, and often disregarding principles of safety and pollution abatement. Most of these nations have only one aim: the immediate returns and revenues for industrialization. This is understandable. But whether such rapid industrialization is a sound policy for the management of the resources of "Spaceship Earth," or whether, instead, the developing nations should carefully evaluate it and beware of the fate to which it might lead, should be a matter of meticulous international assessment and decision making. The developing nations have the opportunity to witness the results of experiments already performed for them by the developed nations, and lessons they can learn *before* the fact. Nevertheless, it is also understandable that petroleum companies operating for these nations should find the lack of restraints less costly, more profitable, less restrictive, and the cause of fewer headaches than similar operations off the shores of developed states. The question must be raised about the responsibility of those organizations for adopting their own self-policing methods and educating the developed nations in the necessity of preserving the environment for their own good and for the ultimate benefit of all mankind.

There is no reason to believe that technologically the industry cannot conduct its business and preserve the ocean ecology in a system of mutual benefit. The ocean has become the focus of man's attention and hope, not merely for its mineral and petroleum resources, but even more for its increasing importance as a source of food, a possible future habitat, and a major source of the earth's weather systems and their life-giving processes. There are other tenants using the continental shelf, and their joint activities need to be mutually compatible.

In a publication entitled *Petroleum, Drilling and Leasing on the Outer Continental Shelf—A Summary* (May 1966, p. 20), the Department of the Interior describes the other tenants of the continental shelf regions as follows:

"One of the singular aspects of the Offshore Louisiana situation is that the oil industry has enjoyed extensive and largely undisturbed use of the area for over a decade, during which time it has put up over a thousand permanent structures which would in varying degrees interfere with other uses of the shelf, the overlying sea, and the air above it. This was possible to do because by and large, and almost fortuitously, no other prospective tenants asserted a significant need to use the area for their own purposes.

This is not to say that operators have not had trouble with merchantmen and fishing boats bumping into their platforms (the frequency is increasing) or by having ships' anchors dragged over their pipelines. These difficulties are of a historic nature and a large body of navigational law and practice has been evolved to enable the traditional users of the sea to share its benefits with the least amount of damage to everyone.

What is unique about the present situation is that of the 200,000-plus square miles of Continental Shelf and sea area adjacent to the United States, that portion off the coast of Louisiana is the only part that has experienced significant drilling activity, and it is also the only part that is not almost completely claim-staked by other users. Around the entire length of the remaining coastline, there is scarcely a square mile that is not being used for some purpose and usually for more than one purpose. The chief tenant is the Department of Defense, but not in every case. There are bombing and gunnery ranges, test and calibration ranges, carrier operating areas, submarine operating areas, torpedo firing ranges, transit lanes, and vast and complicated underwater sound surveillance systems tied to each other and to the shore by a network of cables. On the Atlantic and Pacific Coasts there are also a great many more commercial shipping routes than in the Gulf, and the number of clear days is measurably less. There are commercial cables, oyster beds, and fishing shoals to be considered and a growing number of privately owned submersible craft operating in the relatively shallow waters above the shelf. Moreover, beauty-conscious dwellers along the shore are acutely sensitive to the spectacle of oil rigs working offshore at any point within their range of sight. And because the entire area is already in use, the entry of a new tenant cannot be easily accommodated because of the "domino effect" produced upon adjacent areas. Therefore, as oil activity on the Continental Shelf expands into these areas, oil men and their government lessors are likely to be faced with problems of a kind and dimension they have never really encountered before. Their resolution will take much patient negotiating, and a large measure of tolerance by all parties. The Continental Shelf and the sea and air above it may give the appearance of being spacious and empty, when in fact they are not. Far from being empty, the Shelf deserves to be called our Crowded Frontier."

One point of view is that land resource exploitation deserves first priority; it may be in the ultimate interest of all mankind to pursue the land areas and explore their subsurface thoroughly, leaving the ocean as clean as possible for as long as possible. Until the continents are thoroughly explored and their resources exhausted, according to this view, there seems little justification for stampeding to invade the ocean domain, perhaps to cause irreparable damage to its ecology. In the balance of compatible uses of this environment, the demand and the profits have to be carefully and conscientiously weighed against the damage to ecology and the ultimate cost to reclaim it.

GENERAL POLICY FOR SEABED RESOURCES

The previous section on the technology and economics of offshore exploitation reveals two distinct situations, one involving the hard minerals and the other involving offshore petroleum. For hard minerals there is cautious optimism which calls for a policy of encouragement toward the ocean domain, although the technology and economics may not justify deep-sea mining in the near future, and certain other deterrents may hinder the process of decision making by the mining industry.

In the case of the petroleum industry, the present abundance of its product is reassuring. Underwater technology is advancing at a very rapid pace, pushing the industry into offshore development at increasing costs in most phases of operation. Land resources are still plentiful and relatively less costly; American national security appears in no grave danger, now or in the foreseeable future; and the hazards of offshore operations are becoming a major cause for concern on the national and international levels. There are also numerous other tenants who utilize the ocean domain and demand equity and compatibility in the diverse uses of this environment. All these factors seem to call for a deliberate and cautious program of offshore exploitation, encompassing a carefully protracted advance within an established sequence of priorities and an acceptable framework of jurisdiction on the national and international levels.

This go-slow policy is particularly crucial for the continental shelf because technological development there is progressing at a speed that has already rendered obsolete the definition of jurisdictional limits, legal or otherwise. It is conceded that development will be confined, for some time to come, to the continental shelf areas, and that progress into the deep sea is not imminent. However, the confusion created by the Geneva Conventions, particularly the exploitability clause, invites review; definitive political boundaries are needed for the seaward limit of national jurisdictions. Beyond this limit, the deep-sea areas would then be confirmed as the common domain of the community of nations. Whatever regime is suggested for this international deep-sea domain is subject to legal considerations and international approval, but the issue is not as urgent at this time as is the delineation of national jurisdictions.

6

International Concern

The rapid advances in the acquisition of scientific data, and the spectacular development of technological capabilities to exploit the ocean domain, commercially and militarily, have compelled a general awareness of the potential of the oceans in living and nonliving resources. Countries throughout the world have come to recognize the importance of this domain, and there is at least some indication of a trend toward a policy of leaving it free from national domination. More than two-thirds of the planet earth is at stake, and the theme is to explore and exploit its resources for the benefit of all mankind.

The basic attitude is sound and desirable. How to implement it to the satisfaction of all nations is, however, a complex issue of legal, technical, economic, and political problems.

In October 1965, before the U.N. Economic and Financial Committee, Ambassador James Roosevelt urged cooperation for undersea exploration. "It is not too early for this Committee," said Ambassador Roosevelt, "to start dreaming and thinking exciting thoughts about the role the U.N. can take. In saying this, I am not unaware that this organization has already demonstrated a sensitivity to the fact that no one nation can hope to attack the many problems posed by the ocean and that a large enough attack can be launched only if all the nations cooperate."

In 1966, President Johnson said:

Under no circumstances, we believe, must we ever allow the prospect of rich harvest and mineral wealth to create a new form of colonial competition among the maritime nations. We must be careful to avoid a race to grab and to hold the lands under the high seas. We must insure that the deep seas and the ocean bottoms are, and remain, the legacy of all human beings.[45]

Senator Frank Church carried this reasoning further in 1967 by urging an international agreement to confer title on the United Nations to mineral resources on the ocean floor beyond the continental shelf. The agreement he envisioned would regulate the development of these resources and "might not only remove a coming cause of international friction, but also endow the United Nations with a source for substantial revenue in the future." [46]

The World Peace Through Law Conference, held on July 13, 1967, by 2,500 lawyers from 100 countries, adopted a resolution urging a proclamation declaring that the resources of the high seas beyond the continental shelf appertain to the United Nations. The Conference had two broad objectives. The efficient exploitation of the sea for the benefit of all, including private entrepreneurs, and improvement of the lot of mankind as a whole.

The United Nations, however, had already become involved in ocean affairs, although the main impetus toward internationalization of the seabed began in earnest in 1967, following a proposal by the Malta delegation (discussed under a separate heading) to reserve the seabed for peaceful uses and use its resources for the benefit of all mankind.

ORGANIZATIONS FOR MARINE ACTIVITIES

International bodies and mechanisms for promoting and coordinating marine activities among the participating nations are divided into two major groups: nongovernmental and intergovernmental. In 1968, the United Nations Economic and Social Council submitted a report to the Secretary General which contained a detailed survey of existing mechanisms for the promotion and coordination of marine activities at the international level.[47] The principal nongovernmental organizations were contained within the framework of the International Council of Scientific Unions (ICSU). The intergovernmental organizations were for the most part within the United Nations system, although some tended to maintain considerable independence. A third category included bodies to coordinate the work of international organizations within each of the two groups.

NONGOVERNMENTAL ORGANIZATIONS

The nongovernmental organizations engaged in marine activities can be divided into two groups, the largest of which is the International Council of Scientific Unions (ICSU); the others are more or less informal bodies such as associations of regional extent.

ICSU consists of a number of unions classified according to scientific disciplines, and of several special and scientific committees concerned with interdisciplinary problems. The unions having major interests in marine sciences are:

1. International Union of Geodesy and Geophysics (IUGG)
2. International Union of Biological Sciences (IUBS)
3. International Union of Geological Sciences (IUGS)

ICSU committees concerned with marine sciences are:
1. Scientific Committee on Oceanic Research (SCOR)
2. Scientific Committee on Antarctic Research (SCAR)
3. Special Committee for the International Biological Program (SCIBP)
4. Federation of Astronomical and Geophysical Services (FAGS)
5. Comité International de Géophysique (CIG)

The other nongovernmental international organizations are:
1. International Union for Conservation of Nature and Natural Resources (IUCN)
2. Nordic Council for Marine Biology
3. Congress of Baltic Oceanographers
4. Pacific Science Association (PSA)
5. Mediterranean Association for Marine Biology and Oceanology (MAMBO)

INTERGOVERNMENTAL ORGANIZATIONS

Organizations that promote and coordinate marine-related activities on the intergovernmental level are primarily within the United Nations system. As with the nongovernmental bodies, the non–United Nations intergovernmental bodies are confined mainly to specific regions, and are generally concerned with fisheries and living resources.

The United Nations system and its specialized bodies have always conducted activities in marine affairs, particularly in the scientific and exploratory aspects of oceanography, and in the fisheries and living resources of the oceans. Most of these activities are interrelated, and the U.N. bodies concerned with international programs also receive advice from nongovernmental organizations, particularly ICSU. At the present time, the issue of the seabed is the concern of the Committee on the Seabed, reporting directly to the General Assembly (Figure 11).

One of the more active arms of the United Nations is the U.N. Educational, Scientific and Cultural Organization (UNESCO). The UNESCO program is concerned with stimulating and coordinating basic oceanic research and associated scientific work throughout the world, and with providing technical assistance in oceanography to the developing countries. The work involves large numbers of scientists and experts in the diverse fields of oceanography on an international scale. Marine science programs are conducted by UNESCO's Office of Oceanography, which also serves as secretariat for the Intergovernmental Oceanographic Commission (IOC).

The desire of the participating scientists and the oceanographic community at large in involving the support of governments for worldwide

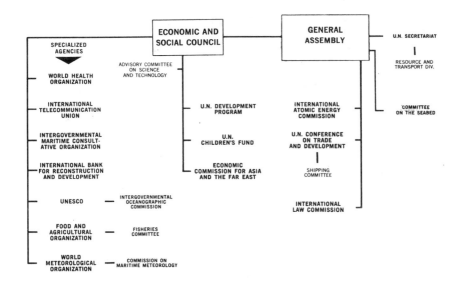

Figure 11. United Nations bodies with responsibilities in the marine sciences. *From* "Marine Science Affairs—A Year of Plans and Progress," the second report of the President to the Congress on marine resources and engineering development, March 1968, p. 24.

cooperation resulted in the establishment of the IOC in 1960. This occurred as a direct outcome of experiences gained during the International Geophysical Year. Since then, the IOC has been actively coordinating major oceanographic expeditions on a global basis, such as the International Indian Ocean Expedition, the Cooperative Investigations of the Mediterranean, and the International Decade of Ocean Exploration.

The Food and Agriculture Organization established in 1961 an Advisory Committee on Marine Resources Research manned by another group of scientists and experts concerned with all aspects of ocean fisheries. The sea and air interface is the target of extensive research and observation networks coordinated by the World Meteorological Organization on a global basis. The International Atomic Energy Agency has an acknowledged competence in the field related to discharge or release of radioactive materials in the sea, and the Inter-governmental Maritime Consultative Organization has an international responsibility to prevent and control oil pollution in the sea through the International Convention for the Prevention of Pollution of the Sea by Oil. IMCO is also concerned with the safety aspects of ships, drill rigs, buoys, and other such platforms at sea.

All United Nations organizations concerned with marine activities coordinate their programs through the Subcommittee on Marine Science and Its Applications, of the Administrative Committee on Coordination, which reports to the Economic and Social Council.

Although intergovernmental organizations other than those within the United Nations system are mainly regional in their extent, one exception, The International Hydrographic Organization, has worldwide interests in (and limited to) hydrography and associated problems of tides and sea level. Its interest in the sea floor is also restricted to its relation to bathymetry. Regional organizations include the International Council for the Exploration of the Sea, which is concerned with the North Atlantic Ocean and adjacent seas, and the International Commission for the Scientific Exploration of the Mediterranean Sea. Both of these organizations cooperate with IOC in coordinating research in their respective areas.

7

U.N. Activities Concerning Seabed Resources

The decade of the 1960s marked a worldwide recognition of the potential of seabed resources. The United Nations was the obvious forum for expressing concern over these resources. The General Assembly and the Economic and Social Council came to recognize that exploration and exploitation of seabed resources should be carried out for the benefit of mankind, particularly toward satisfying the needs of the developing nations. Several resolutions were adopted and decisions made in matters related to seabed resources, with the aim of promoting and facilitating their effective development through coordinated international cooperation.

United Nations activities prior to the 1960s were described earlier, in the discussion on the 1958 Geneva Conventions, particularly concerning the continental shelf. Concerning ocean resources beyond the continental shelf, the Economic and Social Council passed a resolution [1112(XL) on nonagricultural resources. March 7, 1966] requesting the Secretary General:

(*a*) To make a survey of the present state of knowledge of these resources of the sea, beyond the continental shelf, and of the techniques for exploiting these resources, in coordination with those already made by the United Nations Educational, Scientific and Cultural Organization and other specialized agencies and those being prepared;

(*b*) As part of that survey, to attempt to identify those resources now considered to be capable of economic exploitation, especially for the benefit of developing countries;

(*c*) To identify any gaps in available knowledge which merit early attention by virtue of their importance to the development of ocean resources, and of the practicality of their early exploitation.

The General Assembly endorsed Resolution 1112 (XL) in a new resolution [2172 (XXI): Resources of the Sea. December 6, 1966], and further requested the Secretary General:

1. To undertake, in addition to the survey requested by the Economic and Social Council, a comprehensive survey of activities in marine science and technology, including that relating to mineral resources development, undertaken by members of the United Nations family of organizations, various Member States and intergovernmental organizations concerned, and by universities, scientific and technological institutions and other interested organizations;

2. . . . in the light of the above-mentioned survey, to formulate proposals for—
    (a) Ensuring the most effective arrangements for an expanded programme of international cooperation to assist in a better understanding of the marine environment through science and in the exploitation and development of marine resources, with due regard to the conservation of fish stocks;
    (b) Initiating and strengthening marine education and training programmes, bearing in mind the close interrelationship between marine and other sciences;

3. To set up a small group of experts to be selected, as far as possible, from the specialized agencies and intergovernmental organizations concerned, to assist him in the preparation of the comprehensive survey called for in paragraph [1] above and in the formulation of the proposals referred to in paragraph [2] above.

The survey and proposals were to be submitted to the Advisory Committee on the Application of Science and Technology to Development for its comments, and then, together with the comments, to the General Assembly at its twenty-third session (1968), through the Economic and Social Council.

EMERGENCE OF THE MALTA PROPOSAL

While those two reports were being prepared, the Permanent Mission of Malta to the United Nations submitted a *note verbale,* dated August 17, 1967, to the Secretary General, proposing the inclusion in the agenda of the twenty-second session (1967) of the General Assembly an item entitled "Declaration and treaty concerning the reservation exclusively for peaceful purposes of the sea-bed and of the ocean floor, underlying the seas beyond the limits of present national jurisdiction, and the use of their resources in the interest of mankind." [48]

In the memorandum which accompanied the *note verbale,* the Malta proposal pointed out that the seabed and ocean floor beyond the territorial waters and the continental shelves had not yet been appropriated for national use because of their inaccessibility, and because their use for defense purposes or economic development had not been technologically feasible. However, the memorandum recognized the rapid progress in technological developments, particularly by the advanced countries. This progress, it was felt, would cause the seabed to become progressively and competitively subject to national appropriation. National appropriation would, in turn, result in the militarization of the accessible ocean floor through the establishment of fixed military installations and in the exploitation and depletion of resources of immense potential benefit to the world, for the national advantage of the technologically developed countries.

It was, therefore, considered timely—the memorandum continued—to declare the seabed and the ocean floor a "common heritage of mankind." Accordingly, immediate steps should be taken to draft a treaty embodying the following principles:

(*a*) The sea-bed and the ocean floor, underlying the seas beyond the limits of present national jurisdiction, are not subject to national appropriation in any manner whatsoever;

(*b*) The exploration of the sea-bed and of the ocean floor, underlying the seas beyond the limits of present national jurisdiction, shall be undertaken in a manner consistent with the Principles and Purposes of the Charter of the United Nations;

(*c*) The use of the sea-bed and of the ocean floor, . . . and their economic exploitation shall be undertaken with the aim of safeguarding the interests of mankind. The net financial benefits derived from the use and exploitation of the sea-bed and of the ocean floor shall be used primarily to promote the development of poor countries;

(*d*) The sea-bed and the ocean floor, . . . shall be reserved exclusively for peaceful purposes in perpetuity.

The proposed treaty was envisaged to include the creation of an international agency which would assume jurisdiction over the seabed; regulate, supervise and control all activities thereon; and enforce the principles and provisions of the treaty.

Item 92 of the agenda of the twenty-second session of the General Assembly was entitled "Examination of the question of the reservation exclusively for peaceful purposes of the sea-bed and the ocean floor, and the subsoil thereof, underlying the high seas beyond the limits of present national jurisdiction, and the use of their resources in the interests of mankind."

On October 31, 1967, the Secretary General delivered a note (Document A/C.1/952) in connection with this agenda item. He pointed out that the consideration of this item might be facilitated and even sharpened by distinguishing between (*a*) the question of peaceful use, (*b*) the scientific activities, and (*c*) those of resources exploitation. He referred to the studies called for by Resolution 1112 (XL) and Resolution 2172 (XXI) and the progress that had been made in that direction. The Secretary General explained that he had set up a small group of experts to assist him in carrying out the provisions of the resolutions. The group was composed of representatives of the specialized agencies concerned, and of private experts; it held its first meeting in June 1967 at Geneva. The Intergovernmental Oceanographic Commission (IOC) of UNESCO adopted on October 27, 1967, a resolution establishing an IOC working group on the legal questions related to scientific investigations of the ocean.

In connection with these studies, the Secretary General's preliminary work on the tasks outlined in the resolutions led him to the conclusion that there were two major gaps in (*a*) the legal status of the deep-sea resources and (*b*) ways and means of ensuring that the exploitation of these resources would benefit the developing countries. Those gaps had been judged to cause possible delay in the progress of the studies. As to item *b,* the Secretary General suggested the possibility of preparing a more comprehensive report which would include "a study of the legal framework which might be established for the deep-sea resources, the administrative machinery which may be necessary for effective management and control, the possible system of licensing and various possible arrangements for redistributing and/or utilizing the funds which would be derived therefrom, including those earmarked for the benefit of the developing countries."

ORGANIZATION OF THE U.N. SEABED COMMITTEE

An immediate outcome of the Malta proposal was Resolution 2340 (XXII), dated December 18, 1967, by which the General Assembly created an Ad Hoc Committee to Study the Peaceful Uses of the Sea-Bed and the Ocean Floor beyond the Limits of National Jurisdiction. The resolution recognized the extent and speed of developing technology, and that this technology was making the seabed and the ocean floor accessible and exploitable for scientific, economic, military, and other purposes. The Ad Hoc Committee was requested to prepare for the twenty-third session of the General Assembly a study which would include:

(*a*) A survey of the past and present activities of the United Nations, the specialized agencies, the International Atomic Energy Agency and other intergovernmental bodies with regard to the sea-bed and the ocean floor, and of existing international agreements concerning these areas;

(*b*) An account of the scientific, technical, economic, legal, and other aspects of this item;

(*c*) An indication regarding practical means of promoting international co-operation in the exploration, conservation and use of the sea-bed and the ocean floor, and the subsoil thereof, as contemplated in the title of the item, and of their resources, having regard to the views expressed and the suggestions put forward by Member States during the consideration of this item at the twenty-second session of the General Assembly.

The Ad Hoc Committee was composed of 35 members and officers, who were divided into two major groups to consider the requests of the resolution. One was the Economic and Technical Working Group; the other the Legal Working Group. Numerous meetings were held during 1968 in three sessions; it examined the scientific, economic, technical, and legal aspects of the peaceful uses of the sea.

In February 1968, the report "Resources of the Sea," requested by the Economic and Social Council's Resolution 1112 (XL), was submitted for consideration by the Ad Hoc Committee.[49] Part One of the report dealt with the mineral resources of the sea beyond the continental shelf, and Part Two dealt with food resources, excluding fish. The Ad Hoc Committee also considered background papers prepared for it by the Secretariat, the IOC, and other U.N. specialized agencies, and the report on marine science and technology.[50] Its final report reflected emerging conflicts of interest and a heightened awareness of the technical and legal problems associated with exploiting the deep-ocean floor. Earlier anxieties over the seabed resources and expectations of early and large returns from the riches of the seabed became tempered with realism.

When the General Assembly convened in the fall of 1968, it reviewed the Committee report and decided to give the Ad Hoc Committee permanent status. In a series of resolutions, 2467A–D (XXIII) adopted December 21, 1968, the General Assembly established a standing committee —the Committee on the Peaceful Uses of the Sea-Bed and the Ocean Floor beyond the Limits of National Jurisdiction, composed of 42 member states. The Committee was instructed to:

(*a*) Study the elaboration of the legal principles and norms which would promote international co-operation in the exploration and use of the sea-bed and the ocean floor and the subsoil thereof beyond the limits of national jurisdiction and to ensure the exploitation of their resources for the benefit of mankind, and the economic and other requirements which such a regime should satisfy in order to meet the interests of humanity as a whole;

(*b*) Study the ways and means of promoting the exploitation and use of the resources of this area, and of international co-operation to that end, taking into account the foreseeable development of technology and the economic implications of such exploitation and bearing in mind the fact that such exploitation should benefit mankind as a whole;

(*c*) Review the studies carried out in the field of exploration and research in this area and aimed at intensifying international co-operation and stimulating the exchange and widest possible dissemination of scientific knowledge on the subject;

(*d*) Examine proposed measures of co-operation to be adopted by the international community in order to prevent the marine pollution which may result from the exploration and exploitation of the resources of this area.

The rest of this series of resolutions dealt with each of the requests individually: (B) Prevention of Pollution, (C) Study of International Machinery, and (D) Expanded Cooperation and an International Decade of Ocean Exploration, respectively. These resolutions had been cosponsored by the United States, and the International Decade of Ocean Exploration was originally proposed by the United States. However, on the question of establishing international machinery to promote exploration and exploitation of seabed resources and their use, the United States considered the Committee proposal premature and therefore abstained.

Since the establishment of the standing Committee on the Seabed, the United Nations has been actively pursuing ocean affairs in the area of scientific and technological research, disarmament, and the establishment of an international regime for the resources of the seabed. This intensified effort has been matched by activities of the IOC to enable it to serve as focal point for coordinating international marine-science activities, in cooperation with other international organizations, and with U.S. participation.

In 1969, the Seabed Committee established a Legal Subcommittee and an Economic and Technical Subcommittee, which met several times and reported their deliberations and findings to the Committee.

Legal Subcommittee

The Legal Subcommittee was assigned the task of studying the elaboration of legal principles and norms (as described in operative paragraph 2(a) of Resolution 2467A (XXIII)) which would promote international cooperation in the exploration and use of the seabed and the ocean floor, and subsoil thereof, beyond the limits of national jurisdiction and ensure the exploitation of resources for the benefit of mankind, having regard to the economic and other requirements which such a regime should satisfy in order to meet the interests of humanity as a whole. It was also asked to examine the legal implications of all other questions mentioned in the terms of the resolution, and the reports submitted by the Secretary General pursuant to Resolution 2467B, C, & D (XXIII) and 2414 (XXIII).

The subcommittee's deliberations centered around the drafting of a declaration of principles, taking into consideration principles of the Antarctic Treaty, and the concept of "common heritage of mankind." The delegations argued these points at length, but owing to the insufficiency of time the subcommittee decided to postpone consideration of other items until future sessions.

Economic and Technical Subcommittee

The Economic and Technical Subcommittee was asked to consider the following topics:

1. Economic and technical requirements which such a regime as is referred to in operative paragraph 2(a) of Resolution 2467A (XXIII) should satisfy in order to meet the interest of humanity as a whole.

2. Operative paragraph 2(b) of the resolution—to study the ways and means of promoting the exploitation and use of the resources of this area, and of international cooperation to that end, taking into account the foreseeable development of technology and that the economic implications of such exploitation should benefit mankind as a whole.

3. Economic and technical implications of—

    (a) all other questions mentioned in the terms of reference of the Committee as contained in Resolution 2467A (XXIII); and

    (b) the reports submitted by the Secretary General pursuant to Resolutions 2467B, C, & D (XXIII) and 2414 (XXIII).

The subcommittee found that little change had taken place in technological development since the submission of the report on the Resources of the Sea, particularly in the mining techniques. Exploration and exploitation of petroleum, on the other hand, were progressing at an increasing pace. Industry was becoming increasingly aware of the vast mineral deposits contained in the ocean floor, which could in the future become technically and economically exploitable.

The report of the subcommittee recognized the lack of basic documents—geological, topographical and geophysical, etc.—which were needed to identify areas favorable for the occurrence of various minerals and to appraise their potential. The report urged international cooperation in collecting these data and recommended that the developing countries should become more involved in participation in such projects.

International Machinery

The subcommittee considered extensively the report of the Secretary General, which suggested possible functions and forms of international machinery.[51] The functions and powers would include registration, licensing, operation by an international agency, and the settlement of disputes.

One function which international machinery could fulfill would be to provide a system of registration whereby states or other applicants could notify an international body of the activities undertaken or proposed, and of the area in which they would be conducted. The committee found that the main feature of the numerous proposals put forward by governments for licensing was that title or control of seabed resources would be held by the international community, represented by the international authority, which would issue licenses to individual operators. For operations on the seabed, an international body would be established which would exercise its functions in one or a combination of ways: The agency itself might carry out direct exploration and exploitation operations, with its own staff and facilities; it might arrange for others to perform these operations on its behalf by a system of service contracts or possibly by issuing licences; or joint ventures could be undertaken with other bodies, such as government enterprises or international consortia. It was also suggested that international machinery could be established to provide a means for the settlement of disputes arising out of the development of seabed resources.

The proposed functions might be carried out by various forms of international machinery. Possible forms identified in the report included:

1. A secretariat center or unit which might be established within an existing organization, such as the Center for Development Planning, Projections and Policies; the Center for Housing, Building and Planning; and the Center for Industrial Development.

2. A United Nations subsidiary organ, such as the U.N. Conference on Trade and Development (UNCTAD); the U.N. Children's Fund (UNICEF); and the U.N. Relief and Works Agency for Palestine Refugees in the Near East (UNRWA).

3. A United Nations subsidiary organ performing functions under treaties such as the bodies concerned with narcotic drugs and the Office of the U.N. High Commissioner for Refugees (UNHCR).

4. An international organization established by treaty, enjoying an independent legal status, such as the United Nations itself and the specialized agencies.

The Economic and Technical Subcommittee found the report of the Secretary General a useful basis for deliberation, and concluded that of the three functions considered (licensing, registry, and operational agency) the first two had been covered in an overall comprehensive manner. The question of an organization to perform these functions was discussed extensively, but the consensus was that the whole subject of international machinery needed to be considered simultaneously with the legal regime, and that such problems as definitions of the limits of the area and the authority of the agency should be considered further.

The General Assembly then passed resolutions for the continuation of the activities of the Seabed Committee, concentrating on three major issues:

1. Ascertaining member views on convening Law of the Sea Conference to update the Geneva Conventions—Resolution 2574A (XXIV);

2. Requesting the U.N. Seabed Committee to prepare seabed principles and rules for exploitation of seabed resources—Resolution 2574B (XXIV);
3. Requesting a further study on international machinery—Resolution 2574C (XXIV).

Another resolution was passed [2574D (XXIV)] calling for a moratorium on exploitation of seabed resources pending establishment of an international regime.

8

U.S. Participation in International Ocean Activities

The formulation of ocean policy in the United States is a complex process which involves Federal agencies in the executive branch, congressional committees, and other non-Federal and academic organizations. Each one of these bodies contributes a share toward the evolution of U.S. policy, and their individual positions on major issues are not necessarily similar. In order to understand these complexities, it is necessary to identify the organizational structure of the policy-making apparatus and to review the positions taken by the legislative and executive branches of the U.S. Government.

U.S. POLICY APPARATUS FOR SEABED ISSUES

The building of a consensus on marine affairs through studies, expert testimony, and expressions of opinions is a function of several congressional committees. This is usually the formative stage in the process of establishing facts and formulating policy guidelines to assist the executive branch in its tasks. In the executive branch, the policy apparatus included the National Council on Maritime Resources and Engineering Development; the Commission on Marine Science, Engineering, and Resources; the Committee on International Policy in the Marine Environment; and the existing Interagency Law-of-the-Sea Task Force under the Department of State. Outside the Federal Government, assistance is also provided by the National Academy of Sciences and the National Academy of Engineering.

Congressional Committees

In the legislative branch, numerous committees and subcommittees are involved, either directly or indirectly, with ocean-related activities. Committees directly concerned with the outer continental shelf and the international aspects of ocean affairs are the following:

In the House of Representatives:
    Committee on Foreign Affairs
        Subcommittee on International Organizations and Movements
        Subcommittee on National Security Policy and Scientific Developments
        Subcommittee on State Department Organization and Foreign Operations
    Committee on Merchant Marine and Fisheries
        Subcommittee on Merchant Marine
        Subcommittee on Oceanography

In the Senate:
    Committee on Commerce
        Subcommittee on Oceans and Atmosphere
    Committee on Foreign Relations
        Subcommittee on Oceans and International Environment
    Committee on Interior and Insular Affairs
        Special Subcommittee on Outer Continental Shelf

The Senate Subcommittee on Oceans and Atmosphere, chaired by Senator Ernest F. Hollings, was formerly the Subcommittee on Oceanography, which was created as the Special Study on United Nations Suboceanic Land Policy. The Subcommittee on Oceans and International Environment was formerly the Subcommittee on Ocean Space, created also in the 91st Congress to consider the major aspects of the ocean-space issue, including the military, economic, scientific, and legal aspects interacting to form the international issue before the United Nations. It is chaired by Senator Claiborne Pell, and the Special Subcommittee on Outer Continental Shelf was chaired by Senator Lee Metcalf. All three subcommittees were established in 1969 and held hearings on issues related to the United Nations and the seabed.

In matters of scientific and technological nature, advice and assistance for the formulation of policy are provided to the committee staff and members by the staff of the Science Policy Research Division of the Congressional Research Service at the Library of Congress, which was created for that purpose in 1964.

Council on Marine Resources and Engineering Development

In June of 1966, Congress passed the Marine Resources and Engineering Development Act, which became Public Law 89–454 (Appendix 13), establishing policies and objectives for the U.S. effort to develop the Nation's marine resources. It also provided for the establishment of a National Council on Marine Resources and Engineering Development under the chairmanship of the Vice President.

The duties and responsibilities of the Council were outlined in detail in the act, and represented a wide-ranging mandate over the total national program in oceanography. The Council advised and assisted the President in carrying out his responsibilities under the act. These included evaluation of Federal marine-sciences activities, the development of a comprehensive program, the establishment of long-range studies, coordination of a program of international cooperation, and guidance for sea-grant-program policies.

The staff of the Council was composed of specialists in ocean sciences, engineering, national security affairs, economics, foreign affairs, and public administration. It maintained working relations with the Congress, key officials of the Executive Office of the President, Federal and state agencies, industry, the academic community, and professional societies to ensure that considerations affecting all marine-science interests were brought to the attention of the Council.

Commission on Marine Science, Engineering, and Resources

To complement the role of the Council, the act provided for an independent advisory Commission on Marine Sciences, Engineering, and Resources. The Commission was made up of 15 members from the Federal Government and state governments, industry, and laboratories and other marine-science institutions. Four members of Congress served as advisors to the Commission.

The Commission was charged with the responsibility to "make a comprehensive investigation and study of all aspects of marine science in order to recommend an overall plan for an adequate national oceanographic program that will meet the present and future national needs." The findings of the Commission would then be submitted to the President, and the Council would assist the President in evaluating and reviewing the Commission's findings. Thereafter, the Commission would disband, and the Council's authority would be terminated 120 days after the submission of the Commission's report.

The Commission's report, entitled "Our Nation and the Sea", was submitted in January 1969. Some of its recommendations concerning international affairs in the marine environment will be discussed later.

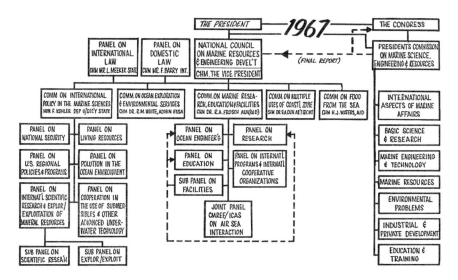

Figure 12. Organization chart showing the relationships within the Federal Government for decision making in marine affairs (1967).

Committee on International Policy in the Marine Environment

As the Marine Council began to coordinate the Federal marine affairs, it created several committees and panels, one of which was the Committee on International Policy in the Marine Environment (Figure 12). This committee was responsible for U.S. foreign policy pertaining to the marine environment; international activities and initiatives pertaining to the marine environment, including cooperation by the United States with other nations and participation in international organizations and meetings. The committee, chaired by the Deputy Under Secretary of State, established a special working group to handle the problem of the U.N. proposals and the U.S. position. This working group consisted of representatives from the Departments of State, Interior, Commerce, Defense, and Transportation, and the National Science Foundation.

Interagency Law-of-the-Sea Task Force

One of the major recommendations presented in the Marine Commission's report called for the establishment of an independent Federal agency —National Oceanic and Atmospheric Agency (NOAA)—to unify the

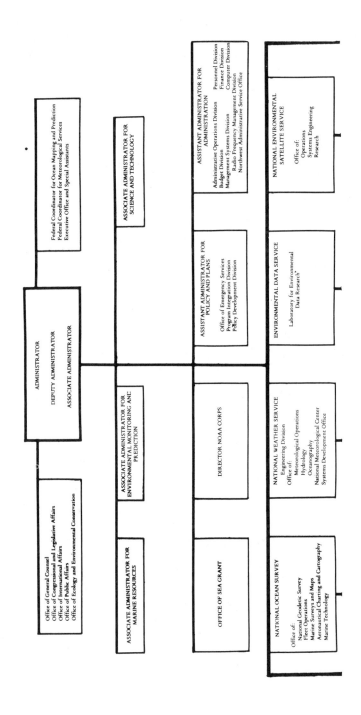

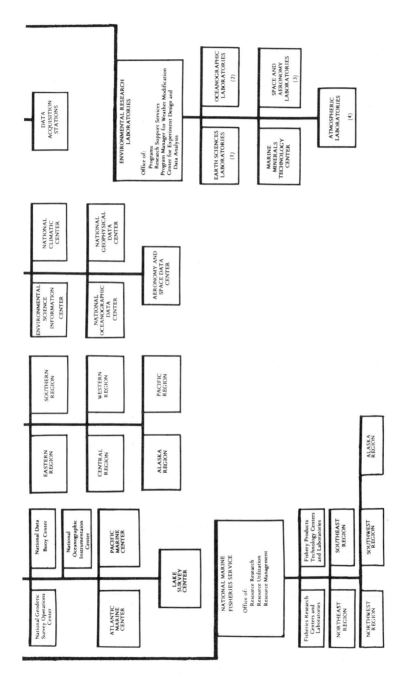

Figure 13. Organization chart showing the components of the National Oceanic and Atmospheric Administration (NOAA), established in the Department of Commerce on October 3, 1970.

national effort in marine affairs, and a non-Federal body called the National Advisory Committee on the Oceans and Atmosphere (NACOA).

While the executive branch was weighing these recommendations, the Congress extended the life of the Council to preserve its coordinating functions. In October 1970, the National Oceanic and Atmospheric Administration (NOAA) was established in the Department of Commerce (Figure 13) but NOAA did not receive as wide a range of functions as the Commission recommended. In October 1971, two months after Congress enacted Public Law 92–125 to establish NACOA, the President appointed 25 members from all sectors concerned with ocean affairs. NACOA would provide advice to the Secretary of Commerce, but would not formulate policy, and is administered by staff from the Department of Commerce.

Although numerous ocean activities remained outside the NOAA, which would require coordination functions similar to those the Council had been performing, the Council was not funded for 1971 and became defunct. Its Committee on International Policy in the Marine Environment had not functioned during 1970, and a new body was formed to take its place.

The new body is now the only policy apparatus responsible for formulating the U.S. position regarding international marine affairs and seabed resources. It is called the Interagency Law-of-the-Sea Task Force, under the chairmanship of the Legal Adviser of the Department of State (now John R. Stevenson). Its members include representatives from the Departments of State, Commerce, Interior, Transportation, and Justice; and the National Science Foundation and the National Security Council.

The National Academies

Outside the Federal structure, the National Academy of Sciences (NAS) and the National Academy of Engineering (NAE), through their committees on oceanography and ocean engineering, respectively, have served as a source of scientific advice to Government agencies on ocean affairs. Toward the end of 1970, the two academies reorganized their ocean-policy structures, raising them to the board level. The National Academy of Sciences' Ocean Affairs Board became administratively lodged in the National Research Council's Division of Earth Sciences, replacing the Committee on Oceanography, and the National Academy of Engineering redesignated its Committee on Ocean Engineering as the NAE Marine Board. The NAS board is concerned primarily with science-related ocean problems, and NAE board with engineering-related problems. Appointees from both boards make up a new ad hoc NAS–NAE ocean affairs planning group to coordinate ocean studies and advisory services of both boards.

The NAS Ocean Science Affairs Board focuses on at least three areas: ocean sciences, ocean resources, and international marine-science-affairs policy. In addition, the board serves as the U.S. Committee for the Scientific Committee on Oceanic Research of the International Council of Scientific Unions.

The NAE Marine Board comprises panels which cover such functional engineering areas as transportation, construction, resource development, and instrumentation. It serves as the U.S. committee for the Engineering Committee on Oceanic Resources, now affiliated with the World Federation of Engineering Organizations.[52]

FORMULATION OF U.S. POLICY FOR THE SEABED

Legislative Concern in the 90th Congress

Reaction in the 90th Congress to the Malta proposal was immediate in the face of imminent and possibly hasty action by the U.N. General Assembly. Almost three dozen resolutions were introduced in the House and the Senate, mostly in opposition to vesting control over the deep-ocean resources in the United Nations. House resolutions were, for the most part, identical, expressing the sense of Congress that any action at that time to vest control of deep-ocean resources in an international body was premature and ill-advised, and that the Congress should memorialize the President to instruct the American representatives to the United Nations to oppose any action to vest in the United Nations control of the resources of the deep sea beyond the continental shelves of the United States.

Hearings were held in the House by the Committee on Foreign Affairs, Subcommittee on International Organizations and Movements, in September and October 1967, and jointly with the Subcommittee on Oceanography of the House Committee on Merchant Marine and Fisheries in June and July 1968.[53] In the Senate, the Committee on Foreign Relations held hearings on Senate resolutions submitted in support of international control.[54]

House Support for the Malta Proposal

Witnesses testifying in the House included several Members of Congress in support of their own resolutions, representatives of Federal agencies directly involved with the U.N. issue, and several representatives of private legal, and industrial organizations.[55] The Subcommittee on International Organizations and Movements addressed itself to the wording of the resolutions, the procedures used in arriving at the U.S. Government's position on the pending legislation and the Malta proposal, the operational marine programs of various U.S. agencies and to the complex legal, political, and economic considerations involved in this legislation.

A number of witnesses who favored the Malta proposal identified advantages that might be derived from international control, such as regulation of the depletion of mineral resources, avoidance of an anarchic rush to claim and exploit subsea resources, reduced danger of marine pollution (through proper international control), reduced threat of a military race to exploit strategic advantages of submarine weapons placement, provision of an independent income for the United Nations, and a general strengthening and maturity in the U.N. itself, through the experience of administering the vast area of the ocean floor.

Proposal for U.N. Marine Resources Agency

All the advantages mentioned above could be provided through a specialized agency like a U.N. Marine Resources Agency. This agency would "hold ownership rights and grant, lease, or use these rights in accordance with the principles of economic efficiency and the well-being of mankind. It should distribute the returns from such exploitation in accordance with the directives issued by the U.N. General Assembly." [56]

Concerning the establishment of an international or U.N. agency, the Department of the Interior pointed out that the result would be something comparable to what already exists in the Food and Agriculture Organization and to some extent in other organizations like UNESCO (see Figure 12). The agency would have responsibility for coordinating exploration and research in the oceans along the same lines as these other organizations; so there would not be anything new and different about it. By analogy with agricultural research, it was suggested that multinational programs tend to disseminate useful results more globally than do single-nation or bilateral research programs.

U.N. agencies have primarily directed their attention to problems of nations which have a low technical capacity of their own to carry on research. The agencies are also able to direct collective attention to the acquisition of information which would help answer international problems of resources management. Those activities do not lessen the need for any nation to carry on research activities in its own interests.

Support in the Senate

Although the majority of House resolutions opposed this approach, the Senate resolutions introduced by Senator Claiborne Pell were strongly in favor of international cooperation, including a "Declaration of Legal Principles Governing Activities of States in the Exploration and Exploitation of Ocean Space." [57]

Hearings were held before the Senate Committee on Foreign Relations on Senate Joint Resolutions 111, 172, and 186.[58] Senate Joint Resolution 111 was identical to the House resolutions opposing the Malta proposal. It was described as overstating the immediacy of the problem by addressing itself to a danger which was not present. The sponsor of the bill, Senator Norris Cotton, had no objection to the United Nations' plowing the ground and preparing the way in this matter of jurisdiction over the riches of the sea. He said he wished Congress to become accessory before the fact and not just after the fact:

> As to the form of my resolution [he said], I would say very frankly to the committee, that the first draft of this resolution was prepared for me by representatives of the National Oceanographic Association. As far as I am concerned, this matter in my resolution of directing the American representatives in the United Nations to oppose action or to take any particular attitude, I think, might well be deleted.[59]

Reasons for Opposition

The objections to any U.N. action stemmed primarily from fears that the United States might be giving away some valuable assets and rights the extent of which were not yet known. A hasty action to relinquish these rights to the United Nations was deemed inadvisable.

Some Members of Congress considered the United Nations unqualified to assume such broad responsibilities.* Scientific exploration, claimed some, could be seriously hampered by a premature definition of political jurisdiction. The issue of national security was also invoked as an objection to any action that might not be compatible with the military programs of the United States. One Member, Representative Paul Rogers of Florida, urged that for exploitation purposes the United States should have the right to occupy the ocean floor to the Mid-Atlantic Ridge and assume the responsibility to defend it.

The objections finally boiled down to the timing of a move to determine where sovereignty lay or to effect a transfer of sovereignty to the United Nations. There were also certain misgivings expressed about the validity of existing definitions of the continental shelf, and a desire to clarify and review these definitions before any final actions were contemplated.

Congressman Alton Lennon, Chairman of the Subcommittee on Oceanography, reminded the Congress that studies were being conducted by the Marine Council and the Commission on Marine Science, Engineering, and Resources on the best solution for controlling the exploitation of

---

* Representative H. R. Gross stated: "Of course, I don't think the United Nations stands for much of anything. It never has and probably never will, and that is one of the reasons why I don't want to see any authority in this matter vested in the United Nations." (In *The United Nations and the Issue of Deep Ocean Resources*, p. 100.)

mineral resources from the continental shelves. Consequently, he said, it was in the national interest to wait for these studies to be completed "as keenest minds available in international law and marine science study all aspects of this complex problem in the hope that an equitable solution can be resolved for all nations." [60]

Legislative Concern in the 91st Congress

As the 91st Congress convened, it had before it "Our Nation and the Sea"—the report of the Commission on Marine Science, Engineering, and Resources presented to the President and to Congress on January 9, 1969.[61] This report recognized the inadequacy of the present framework for the continental shelf and the seabed beyond. It recommended a precise definition of the continental shelf—a limit of each coastal nation to be fixed at the 200-meter isobath, or 50 nautical miles from the baseline for measuring the breadth of the territorial sea, whichever alternative gave it the greater area. For the seabed beyond these limits, the Commission proposed a new international legal-political framework for exploration and exploitation of the mineral resources underlying the deep seas. It proposed further the establishment of an International Registry Authority, and an intermediate zone between the limits of the continental shelf and the deep-sea area. The intermediate zone would begin at the 200-meter isobath (or 50 nautical miles from the coast) seaward to the 2,500-meter isobath (or 100 nautical miles, whichever was farther from shore). The report proposed policy guidelines and goals for the United States to follow in considering the needs to implement these recommendations.

The recommendations, the activities of nations on the U.N. Seabed Committee, and an executive-branch proposal concerning the seabed, submitted on August 3, 1970, raised a series of questions which became the focus of attention by the 91st Congress. What were the limits of the continental shelf? Should the limit be geological or legal? Should it be based on considerations of equity, security, or economic advantage? How much did the United States stand to lose by the creation of an international regime? Was a new Law of the Sea Conference necessary? Should the states have a narrow or a wide continental shelf? For areas beyond the continental shelf, what sort of an international regime would be best? What principles should be adopted? What kind of international machinery should be established? How did all these aspects affect the economy and national security of the United States?

A series of hearings in the Senate sought the answers to those questions. The Committee on Foreign Relations Subcommittee on Ocean Space, chaired by Senator Claiborne Pell, heard testimony on his Senate Resolution 33, which proposed basic principles to govern the development and utilization of the ocean-space environment.[62]

The Committee on Commerce also held hearings through its Special Study on United Nations Suboceanic Lands Policy, chaired by Senator Ernest F. Hollings. This study group was formed in July 1969 for "the purpose of considering the policy which the United States should advocate within the United Nations when that organization considers the ground rules which should apply to those nations which desire to exploit the resources of the deep oceans." [63] The hearings were intended to enable the committee to make recommendations to Senator Pell's subcommittee and to the Senate. Senator Pell's subcommittee members and members of the newly formed Special Subcommittee on Outer Continental Shelf, Committee on Interior and Insular Affairs, were invited to participate. Similar participation took place when the Special Subcommittee on Outer Continental Shelf held its hearings,[64] chaired by Senator Lee Metcalf.

The intent of the three sets of hearings was similar, and most of the witnesses testified on the same subject before more than one subcommittee. The Metcalf subcommittee, in particular, compiled a voluminous record of statements by representatives of the Departments of State, Defense, Commerce, Interior, and Transportation; the scientific and industrial sectors; and numerous distinguished international lawyers. The subject matter included legal and political aspects of the definition of the continental-shelf boundaries, the economic and conservation aspects related to alternative boundary locations, comments on the moratorium resolution and the interim policy for the seabed, and the position of the executive branch regarding all these aspects. The hearings were followed by a painstaking analysis of findings in a subcommittee report which is discussed below. The hearings of Senator Hollings' Special Study on United Nations Suboceanic Lands Policy and those of Senator Pell's Subcommittee on Ocean Space were less comprehensive and did not result in position papers. Senator Pell took the same position as he had during the 90th Congress and in testimony before the Metcalf subcommittee.

Senator Pell's Proposals

In his Senate Resolution 33, Senator Pell submitted a "declaration of legal principles governing activities of states in the exploration and exploitation of ocean space." These principles called for the use of the seabed and subsoil for peaceful purposes only, under licenses issued by a technically competent licensing authority to be designated by the United Nations; regulations on the disposal of radioactive waste material in ocean space; the establishment of a Sea Guard under the control of the U.N. Security Council; and a definition of limits of the continental shelves.

Although Senator Pell, in his testimony before the Metcalf subcommittee, upheld the recommendations made by the Marine Commission, the limits of the continental shelf proposed in his resolution differed from those proposed by the Commission. He preferred the 550 meter isobath or a distance of 50 nautical miles from the baselines used to measure the breadth of the territorial sea, whichever gave the coastal state a greater area offshore for purposes of mineral resources exploitation. "I selected the 550-meter figure," he testified, "on the basis that the edge of the outer continental shelf is not known to occur at any greater depth." [65] In effect this isobath encompasses the topographic configuration of most of the world's shelves, to the greatest depth, rather than to the average depth of 200 meters. Mr. Pell made a distinction between the "continental terrace" and the "continental shelf," that differed from the position of the oil industry of equating the two.

And here, Mr. Chairman, I must emphasize the shell game in which, I believe, the oil industry has been engaged in the past few years: when you and I went to school, continental shelf meant that portion of the submerged continental land mass that is in relatively shallow water and terminates at the beginning of the continental slope. The oil industry has thrown up a smoke screen by trying to equate the continental shelf with the continental terrace concept, which includes the slope.[66]

On how wide the national jurisdiction of a state should be offshore, Mr. Pell favored the narrowest possible zone. If the United States claimed a certain width, he argued, it should be assumed that other nations would do likewise. "Thus the larger the offshore zone we contemplate bringing under our national jurisdiction means that on balance we are closing off a much larger zone worldwide, assuming as we must that other states would be entitled to claim a similar area." [67]

Later in 1970, Senator Pell reiterated his support for the early achievement of an international legal order governing the ocean floor in his comments on the President's proposal for a seabed regime:

I bring up this point, Mr. President, merely to show how an issue as important as nuclear arms control can suffer because of the chaos nations have made of the law of sea; and here I should point out that in earlier years the United States has been just as guilty in helping to create this sad state of affairs as Chile and Ecuador or, more recently, Brazil and Canada.

But we now understand the error of our ways, and I am convinced that the policy initiatives which this administration has taken in trying to bring about a meaningful legal order of the oceans deserve the full support of every Member of this Chamber.[68]

Position of the Subcommittee on Outer Continental Shelf

A systematic analysis of the hearings was presented in the Metcalf subcommittee's report to the Committee on Interior and Insular Affairs.[69] The subcommittee considered the 1958 Geneva Convention validly operative, and saw no need to convene another Law of the Sea Conference. It also considered the geological interpretation of the continental margin as making that portion of the seabed off the United States essentially property of the United States. The subcommittee further indicated its preference for the exploitability clause in the Convention which expands the limits of the shelf in keeping with the technological capability to exploit in deeper waters —the principle of expanding boundaries.

The subcommittee first asserted its jurisdiction over policy issues affecting the continental shelf of the United States. It cited the Interior Committee's work on the Outer Continental Shelf Lands Act and the Submerged Lands Act of 1953, giving it the responsibility for legislative oversight of operations under that law and any subsequent amendments of these acts. Assuming that the shelf was an integral part of the continental United States, and interpreting the Constitution (Art. IV, sec. 3, cl. 2) concerning this issue, the subcommittee declared that any modification of the property rights of the United States created by or reaffirmed in these acts would require an act of Congress.

The subcommittee adopted the interpretation agreed upon by the American Branch of the International Law Association that "right [i.e., sovereignty] under the 1958 Geneva Convention on the Continental Shelf [should] extend to the limit of exploitability existing at any given time within an ultimate limit of adjacency which would encompass the entire continental margin." [70] The subcommittee supported the objectives calling for a stable system of law applying to the deep seabed and assurance of the continued freedom for scientific research; however, it also held that:

> . . . undisputed access to the vast energy resource [oil, in particular] located on the U.S. continental margin is of paramount importance. Oil is a strategic material which is absolutely essential to fuel our industrial machine and thereby sustain a sound economy. [71]

As to the boundary limits of the shelf, the subcommittee argued against a narrow shelf and the premise of upholding the freedom of the seas through larger internationally controlled ocean space. It upheld the Geneva Convention as "sufficiently precise as to permit a positive, reliable, and adequate interpretation of the breadth of the legal shelf." It also interpreted the Convention to hold that "the sovereign rights of coastal nations to explore and exploit their legal Continental Shelf extend to the limit of exploitability existing at any given time within an ultimate limit of adjacency which encompasses the entire continental margin." [72] Furthermore, it contended that the drafters of the Convention had limited the jurisdictional claims to the natural resources of the submerged land in order to preclude any abrogation of freedom of the high seas. Hence, the expanding boundary concept was "consistent with the intent of the Convention's drafters as it is an additional means of prohibiting jurisdictional claims not related to the exploration and exploitation of the natural resources of the submerged land continent." [73]

The subcommittee endorsed the general features of the new policy statement submitted by the President in May 1970, calling for a seabed treaty, an international authority, and the renunciation of sovereign rights of all nations beyond the 200-meter isobath. However, it had objections to the renunciation of "the heart of our sovereign rights," particularly relative to the continental margin.

For the United States [asserted the subcommittee], or any other law-abiding nation, to offer to renounce its inherent sovereign rights to the mineral estate of its continental margin in the hope that these few recalcitrant nations would mend their ways and begin to adhere to the freedom of the seas doctrine is like offering to pay ransom to bandits in order to encourage them to stop stealing. When bandits receive ransom, they only grab for more. Thus, to renounce what constitutes the heart of our sovereign rights in response to illegal demands by a handful of nations can only encourage greater violation of the freedom of the seas doctrine.[74]

On resources of the seabed beyond the continental margin, the subcommittee shared with the President the desire that such ocean resources be used rationally and equitably for the benefit of mankind. However, prior to the adoption of a seabed treaty, cautioned the subcommittee, "the U.S. Government should provide measures designed to insure protection of investors who desire to exercise present high seas rights to explore and exploit the wealth of the deep seabed beyond the limits of the submerged land continent." [75]

The subcommittee concluded that the major tasks to be considered in the 92d Congress were:

1. A continuing extensive review of the working paper introduced by the U.S. delegation at the August [1970] session of the United Nations Seabed Committee with a view toward seeking modifications of it to conform to our interpretation of the President's intent and with our recommendations outlined above.

2. An investigation of the special problem of an interim policy which would insure continued exploration and exploitation of the natural resources of our continental margin under present law; and would establish appropriate protection for investments related to mineral recovery by U.S. nationals in areas of the deep seabed beyond the limits of exclusive national jurisdiction.[76]

Position of the Executive Branch

In the section of this study that discussed the continental shelf, it was shown that U.S. policy on ocean resources began in earnest with the Truman Proclamation of 1945. The proclamation was designed primarily to provide a policy and legal framework for regulating offshore operations of the U.S. petroleum industry. When viewed in the perspective of international legal concepts and the world's technological capabilities at that time, the Truman Proclamation might have been considered unnecessary. Ocean technology was then almost exclusively possessed by the United States; no other nation had the technical capability to exploit the resources of the U.S. continental shelf. The proclamation had the effect of stimulating proclamations by other countries, such as the Declaration of Santiago in 1952, whereby national sovereignty and jurisdiction were extended out to 200 miles offshore. Thus, the Truman Proclamation could be taken as the beginning of legal chaos in international maritime affairs, which has persisted to date despite the efforts made at the 1959 Geneva Conventions.

Experience pointed to the conclusion that unilateral action—perhaps accompanied by a scramble to stake out national claims to the "riches of the sea"—has the effect of eroding the freedom of the seas and proves a practical detriment to the world community at large. President Johnson, in his Washington Navy Yard speech had warned of precisely this consequence of the "race to grab and to hold" and had called instead for preservation of these resources as a "legacy of all human beings." [77]

This statement set the course for the position taken by the United States during deliberations following the Malta proposal. Testifying before the 90th Congress, spokesmen of the Departments of Interior and State had generally affirmed that in dealing with areas beyond the jurisdiction of national states, that is, beyond the continental shelf, regardless of its definition, the United Nations should, logically, be concerned with the subject. The United States was in the process of developing its own policy objectives through the Marine Resources Act of 1966; consequently, no support was contemplated for the treaty envisaged by Malta.

Fears of hasty action were allayed by the State Department's expression of doubt that the General Assembly could get very far with a proposal of this specificity on such short notice. It was pointed out that there would have to be a process of study through committees and specialists, and the deliberative process in the United Nations tends to be lengthy.

On September 21, 1967, the U.S. Ambassador to the United Nations supported the inscription of the Malta proposal on the agenda of the U.N. General Assembly and asserted that the United Nations was in a position to assume leadership in enlisting the peaceful cooperation of all nations in developing the world's oceans and their resources.

Following establishment of the Ad Hoc Committee in December 1967, the United States participated in its deliberations and on June 28, 1968, submitted a number of proposals (Appendix 14), including a draft resolution containing (*a*) a declaration of principles on the use of the deep-ocean floor; (*b*) a draft resolution referring to the Eighteen-Nation Disarmament

Committee (ENDC) the question of arms limitations on the seabed and ocean floor with a view to defining those factors vital to a workable, verifiable, and effective international agreement which would prevent the use of this new environment for the emplacement of weapons of mass destruction; and (*c*) a suggestion to establish international marine preserves.

The United States also supported a less extensive declaration of principles submitted by a number of delegations. These principles differed from previous U.S. positions in that the United States came to recognize the "interest of the international community in the development of deep ocean resources," and the "dedication as feasible and practicable of a portion of the value of the resources recovered from the deep ocean floor to international community purposes." Draft resolution C proposed the International Decade of Ocean Exploration (IDOE) for broadening and accelerating investigations of the oceans, and for strengthening international cooperation. IDOE was adopted by the General Assembly as part of the long-term and expanded program of worldwide exploration of the oceans and their resources under the direction of UNESCO's Intergovernmental Oceanographic Commission.

Seabed Disarmament Treaty

In its opening paragraph on ocean science and technology and national security, the President's Science Advisory Committee stated:

> The most urgent aspect of Federal involvement in ocean science and technology for the next 5 to 10 years relates to national security in the narrow, strictly military sense. The U.S. Navy, which has responsibility for essentially all our defense efforts involving the ocean environment, will have increasing need for specialized oceanographic data for specific devices being developed or improved and will continue to require better understanding of characteristics of the ocean environment in which it operates.[78]

Although this statement referred to needs in support of specific projects, it also reflected the need for the U.S. Navy to explore the oceans throughout the world and not merely in the coastal areas of the United States. This need, coupled with the military presence required in numerous parts of the oceans, formed the basic justification for freedom on the high seas, and for the privilege of approaching as close as possible the coasts of other nations. The military view has been, and continues to be, that any extension of territorial seas should be kept to a minimum, that sovereignty over the continental shelves, regardless of their boundaries, should be closely limited, and that the air space above the high seas should remain free. In the words of Dr. Robert A. Frosch, Assistant Secretary of the Navy for Research and Development:

The security of the United States rests in part on the Navy's use of the high seas, and we would like to see the use and legal coverage of the high seas develop in such a way as not to impede this portion of our security unnecessarily.[79]

With this attitude and background, the United States began to evaluate a draft treaty submitted by the U.S.S.R. on March 18, 1969, providing for prohibition of the emplacement on the seabed and the ocean floor and the subsoil thereof of objects with nuclear weapons or any other weapons of mass destruction, and the establishment of military bases, structures, installations, and the like, beyond the 12-mile zone. The measure appeared to call for total disarmament of the seabed, the Soviet Union having equated the uses of the seabed for "peaceful uses" with "nonmilitary purposes," by analogy with the provisions of the Antarctic Treaty of 1959.

The United States considered the proposed complete demilitarization "unworkable and probably harmful." The U.S. representative pointed out that defense against submarines involved placing warning systems on the seabed, and that military personnel participated in scientific research in that environment. On May 22, 1969, the United States countered with its own version of a seabed treaty, prohibiting the emplacement of *fixed* nuclear weapons or other weapons of mass destruction or associated *fixed* launching platforms on, within, or beneath the seabed and ocean floor.

In presenting the draft treaty, the U.S. representative pointed out that the 3-mile territorial sea would leave a larger area subject to prohibition than the 12-mile zone proposed by the Soviet Union. However, the minor disagreements were not insurmountable. On July 3, President Nixon sent a message to the Disarmament Committee stating that it should not be impossible to find common ground between the United States and the Soviet Union in spite of these differences, and that the goal should be to present a sound proposal to the United Nations.

An acceptable proposal came after a full discussion of the two drafts in the form of a joint draft treaty submitted by the United States and the Soviet Union on October 7, 1969. The draft was a compromise between the positions of the two major powers. It provided that:

> The states parties to this treaty undertake not to emplant or emplace on the seabed and the ocean floor and in the subsoil thereof beyond the maximum contiguous zone provided for in the 1958 Geneva Convention on the Territorial Sea and the Contiguous Zone any objects with nuclear weapons or any other types of weapons of mass destruction, as well as structures, launching installations or any other facilities specifically designed for storing, testing or using such weapons.

Later in October, an amended version was submitted at the Geneva Conference of the Committee on Disarmament (CCD, formerly known as the Eighteen-Nation Disarmament Committee—ENDC), now having 26 member nations. This draft was referred by the General Assembly back to the CCD as the "Draft Treaty on the Prohibition of the Emplacement of Nuclear Weapons and other Weapons of Mass Destruction on the Seabed and Ocean Floor and Subsoil Thereof."

Except for minor changes, the definition of the scope of the prohibition remained unaltered in the revised version submitted on April 23, 1970. Strong pressure was applied by the nonaligned nations, which set out amendments resulting in the adoption in the September 1, 1970, revision of a separate article reading:

The parties to this treaty undertake to continue negotiations in good faith concerning further measures in the field of disarmament for the prevention of an arms race on the sea-bed, the ocean floor, and the subsoil thereof.

On December 7, 1970, the General Assembly of the United Nations finally recommended the treaty. On February 12, 1971, ceremonies were held simultaneously in Washington, Moscow, and London; and more than 80 nations have so far signed the treaty (Appendix 15). In the United States, it gained Senate consent for ratification by a vote of 83 to 0 on February 15, 1972.

Seabed Regime

On May 23, 1970, President Nixon released an important policy statement on the seabed (see Appendix 16). He recognized the speed with which modern underwater technology was advancing, and that the prevailing law of the sea was in need of being reshaped and updated to meet the needs of modern technology. He therefore proposed the convening of a new conference on the law of the sea, and consideration of international machinery for authorizing exploitation of seabed resources. The President proposed that "all nations adopt as soon as possible a treaty under which they would renounce all national claims over the natural resources of the seabed beyond the point where the high seas reach a depth of 200 meters (218.8 yards), and would agree to regard these resources as the common heritage of mankind." The regime proposed for the exploitation of seabed resources would provide for the collection of substantial mineral royalties to be used for international community purposes, particularly economic assistance to developing countries. It would also establish rules and regulations for protecting the ocean environment and for safeguarding the investments necessary for exploitation, and a mechanism for the settlement of disputes.

To accomplish these goals, the President proposed two types of international machinery:

First, I propose that coastal nations act as trustees for the international community in an international trusteeship zone consisting of the continental margins beyond a depth of 200 meters off their coasts [Figure 14]. In return, each coastal state would receive a share of the international revenues from the zone in which it acts as trustee and could impose additional taxes if these were deemed desirable.

As a second step, agreed international machinery would authorize and regulate exploration and use of seabed resources beyond the continental margins.[80]

In the meantime, an interim policy was proposed for all nations to join the United States in seeing to it that all permits for exploration and exploitation of the seabed beyond 200 meters be issued subject to approval under the international regime to be agreed upon.

In June 1970, the Committee on Oceanography of the National Academy of Sciences-National Research Council recommended that the United States consider opening ocean waters subject to U.S. jurisdiction to

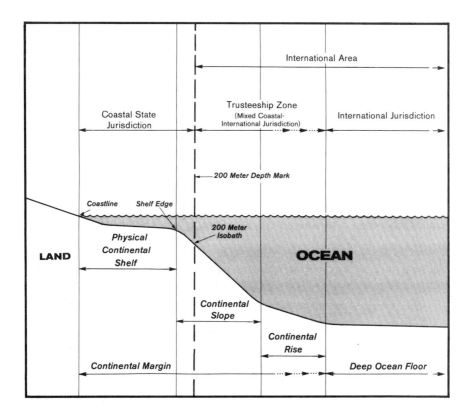

Figure 14. Schematic representation of the continental margin and the seabed, showing zones and boundaries proposed by the United States. Not to scale; compare with Figure 1.

scientific research by foreign nations as a means of encouraging other countries to ease their own restrictions. The resolution called for maintaining appropriate and adequate safeguards for national security, but without requiring researchers to obtain a permit. This policy would not apply to internal waters.[81]

As indicated earlier, hearings were held in the Senate concerning the provisions of the President's proposal, which was formally introduced on August 3, 1970, as the draft United Nations Convention on the International Seabed Area (Appendix 16). On the final day (August 28, 1970) of the session of the Seabed Committee in Geneva, the U.S. Representative commented on the President's proposal:

> When President Nixon made the difficult political decisions inherent in his May 23 announcement and in our draft convention he placed great importance on international community interests. We, as a party to the 1958 Geneva Continental Shelf Convention, could have relied on the exploitability test to extend our boundary unilaterally. We felt, however, that in view of the uncertainties surrounding seabed boundaries, and in light of the great opportunity the international community now has to rectify the inequities of the law of the sea, it would be better for states to renounce under a treaty all national claims beyond the 200-meter isobath, leaving the international seabed area as the widest area possible. By this move we could wipe the slate clean and, in essence, rethink the proper relationship of international community interests to those of coastal states.[82]

The working paper detailed the basic principles concerning mineral resources, living resources, protection of the marine environment, life, and property, and the establishment of an International Seabed Resource Authority to manage the resources, safeguard the investments, and settle conflicts and disputes. During the 25th session of the United Nations General Assembly, these principles were considered, culminating in the passage on December 1, 1970, of two resolutions, one establishing a time and calling for convening in 1973 a new conference on the law of the sea. The other adopted a set of principles in a declaration setting forth the ground rules for ocean-resources management and scientific research.

Reaction to the U.S. proposal was generally not very favorable. In Geneva, the 1971 deliberations by the preparatory group of the U.N. Seabed Committee revealed ominous and discouraging signs of disagreement which may put the conference off beyond 1973. In the United States, as mentioned earlier, considerable opposition has been voiced in Congress as well as by the petroleum, mining, and fisheries industries.

Although fisheries representatives were at Geneva, they were merely observers and performed no advisory functions to the U.S. delegation. In letters to the President and members of Congress, spokesmen for some fisheries claimed that the U.S. delegation showed a lack of concern for the domestic fisheries and that U.S. fisheries were to be sacrificed to achieve other goals. The goals alluded to include the desire of the Department of State to maintain good foreign relations, and the expressed desire of the Department of the Navy to maintain a state of freedom of the seas, with the narrowest possible territorial limits.

Of course, the U.S. delegation cannot include those not officially working for the Federal Government. However, in response to the objections raised, the Department of State established an Advisory Committee for consultation with the Interagency Law-of-the-Sea Task Force. The Advisory Committee includes some 60 representatives from all non-Government sectors concerned with ocean affairs, and provides the Task Force with additional input toward the formulation of ocean policy.

The Senate Committee on Interior and Insular Affairs also continued active interest in the Law-of-the-Sea Conference and assigned Charles F. Cook, Jr., minority counsel, and Merrill W. Englund, administrative assistant to Senator Lee Metcalf, as observers to the July–August 1971 session of the U.N. Seabed Committee meeting in Geneva. The two congressional aides presented the results of their observations in a report to Senator Henry Jackson, Chairman of the committee. They reported that, although the U.S.S.R. supported the U.S. "free transit" proposal, the developing countries "brutally attacked" it and preferred adherence to the traditionally recognized "innocent passage" doctrine. They expressed fear, similar to that of the fisheries industry, that the Defense Department "might urge the Administration to abandon its deep seabed mining objectives . . . as a trade-off . . . in favor of the Defense Department–sponsored free transit proposal." [83]

The report recommended legislation aimed at "reinforcing U.S. rights to mine the deep seabed, encouraging continuation of U.S. leadership in deep sea technology and providing a climate conducive to U.S. investment in deep seabed exploration and exploitation." It also suggested that the Administration review carefully its Draft Seabed Treaty "with an eye toward proposing a new working paper more in consonance with the emerging international consensus." [84]

With this opposition in the background, it is too early to predict what success the present U.S. proposal will achieve. Even if the United Nations succeeds in framing a generally acceptable treaty for the seabed, the final decision in the United States will be subject to approval by the Senate. Some alternative plan may be necessary in the event of an impasse between the executive and legislative branches of Government. What form this alternative might take is difficult to conjecture at this time.

What the proposal, resolutions, and discussions left unresolved was the limit of national jurisdiction. The United States came to look favorably on the 12-mile territorial limit as a probable goal, and several nations appeared receptive to the idea. The Latin American nations clung to their 200-mile limits, and agreements on these limits in future debate may prove hard to reach.

9

Role of Science and Technology in Seabed Diplomacy

As science and technology have reached and affected remote regions of the world, scientists and engineers have begun to play an increasingly significant role in diplomacy. Long the purview of international lawyers and diplomats, ocean sovereignty has been discovered to possess important technical aspects as well. In the field of oceanography, the jurisdictional solutions to man's problems have been essentially an attempt to reconcile man-made laws with the laws of nature. Oftentimes, these two sets of laws have proved incompatible, and the need for knowing and understanding the scientific aspects of the ocean environment has become an obvious prerequisite for successful adjudication among nations.

Understanding of all aspects of the marine environment has also been a major requirement for the proper conduct of naval operations. These operations figure prominently in matters of national security and the formulation of foreign policy, particularly where global commitments are concerned. Since World War II, the outlook toward the use of the oceans for military purposes has assumed progressively larger dimensions. Military strategy has evolved along lines determined largely by developments in technology and by policy goals for internal security and global politics. As one analyst observed:

If asked, an oceanic strategist would tell the President that in order to pursue his policy of nuclear sufficiency and at the same time deter World War III, a blue water oceanic option is the only option for deterrence or defense during the next six years and in the first decades of our Nation's third century.[85]

MILITARY TECHNOLOGY AND OCEAN STRATEGY

The days of bombers and strategic air strikes followed the development of nuclear fusion weapons in 1954, and the threat of Communist expansion.

After 1957, missiles replaced bombers for strategic deterrence, resulting in the development of land-based systems of intercontinental ballistic missiles, and submarine-based nuclear missiles. Military technological breakthroughs continued to reshape military strategy in the 1960s. Improved accuracy and longer range and larger delivery systems permitted the production of large numbers of sophisticated missiles such as the Minuteman and the Polaris submarine-launched system. The development of reconnaissance satellites permitted both the United States and the Soviet Union to maintain surveillance on each other's land-based systems.

Land-based systems (Minuteman and SS–9, for example), however, have their vulnerability: They can be detected by reconnaissance satellites and other means; accuracy of attacking missiles is advancing so fast that it can be measured in a fraction of a mile; MIRV systems can deliver an overwhelming load of warheads, disproportionate to land-based missiles, rendering them increasingly vulnerable.

The shift in strategy toward the ocean environment, therefore, has become obvious and necessary. Numerous analysts have expounded the advantages of undersea weapons systems as being less targetable than their land-based counterparts. The most obvious advantages are their mobility, concealability, and survivability following a sudden nuclear attack. Furthermore, their range throughout ocean space has an added safety factor in deploying away from populated areas.

The absorption of water with respect to light, high-energy particles, electromagnetic radiation, heat and other known forms of energy is such that, except for acoustic radiation, none of the mechanisms postulated has a detection range potential which is significant when compared with the vast areas available in the ocean. The ultimate test in this regard is the ability of the submersible to blend with and be masked by the environment. At near zero speed this ought to be quite modest, and if, for example, power is supplied by fuel cell, the machinery associated with it should be extremely quiet. Drifting in the current, at great depth or at low speeds, the hydrodynamic wake would be insignificant. A further aid would be the capability to move very close to the bottom, rendering the submersible difficult to detect by long-range, active sonar. Ultimately, the underseas weapons systems could develop into something akin to a manned, on-the-bottom, slowly mobile mine.*

For the United States, ocean advantages are enhanced by worldwide interests which have been conventionally served by land bases overseas.

* John P. Craven, "Ocean Technology and Submarine Warfare" in *Implications of Military Technology into the 1970s*, Adelphi Paper No. 46, London, Institute for Strategic Studies, March 1968, pp. 38–46. Dr. Craven was for some years chief scientist of the Special Projects Office of the Department of the Navy, which developed the Polaris system.

These bases have been dwindling in number, and those left are subject to political uncertainties. The 2nd Fleet in the Atlantic, the 6th Fleet in the Mediterranean, and the 7th Fleet in the Pacific are affected by basing problems. Floating support can be maintained similar to the system which largely aids the operation of the 6th Fleet. Objections to this type of support include its high cost and vulnerability. Many alternatives have been envisaged for overseas bases, such as floating platforms of large dimensions, derived from the technology of offshore exploration and exploitation.

Very great changes are expected in naval capabilities as a result of technological developments in submarine warfare. At the start of 1972 the Navy had completed financing the construction of 12 nuclear-powered high-speed attack submarines, and funds are requested for 6 more in the defense budget of fiscal year 1973. The design of these 688-class submarines provides them with the capability to seek and destroy enemy submarines as well as surface ships. The Undersea Long-range Missile System (ULMS) is envisioned as the key element in the U.S. nuclear deterrent arsenal beyond the 1980s. The President has ordered the Department of Defense to accelerate the development of this $15 billion submarine missile system. More than 30 missile-firing submarines are now being armed with the new Poseidon missile. By comparison, the missile designed to arm the ULMS submarine will have a striking range of 6,000 miles, twice that of the Poseidon.

A few years ago, Gordon J. F. MacDonald, then executive vice president of the Institute for Defense Analyses, envisioned for the 1970s that:

> A nation could control the surface of the oceans without having a single ship. The required system would involve satellites equipped with a variety of sensors that would maintain coverage of the world's oceans. Satellites would relay the information to a central computer system which would then target the land-based missiles on ships to be destroyed. The missiles would then be equipped with terminal guidance or be under direct control of the satellite and land-based computer systems. While it is most unlikely that any nation would adopt such a strategy, this example illustrates the fact that naval posture may change radically in the future.[86]

MacDonald also postulates future placement of missiles as large as the Polaris, or larger, on a relatively shallow shelf floor in a barge system that could be moved occasionally to prevent its detection. Another possibility would be mobile ocean-bottom systems which crawl or creep on the seabed. The technology and engineering requirements for manning, maintaining, and servicing these installations would not differ from those used in offshore mineral exploration and exploitation. In fact, even if bottom installations were not militarily desirable (underwater mobility being the key advantage), the thrust into deeper waters of the continental shelf by the petroleum industry might eventually require some kind of protection by the United States, and the Navy might be called on to provide it.

In shaping U.S. policy for the disarmament of the seabed, the effect of technology was much in evidence. The banning of *fixed* bottom installations did not pose any dangers, particularly when the United States had come to realize the importance of mobility for its underwater deterrent systems. In testimony before Senator Pell's Subcommittee on Ocean Space, the following exchange took place between Senator Pell and Dr. Robert W. Morse, president of Case Western Reserve University and formerly Assistant Secretary of the Navy for Research and Development:

> Senator PELL. Do you have any concern about moving in terms of prohibiting mobile weapons systems from operating on the seabed?
>
> Dr. MORSE. No; I do not really—otherwise I think we may end up banning things that do not have any military use and certainly we can get widespread agreement on that. One has to remember that the great advantage of deploying a weapons system at sea is mobility, and that if one bans only fixed nuclear weapons systems at sea he may well be banning something that doesn't have any value anyway.[87]

Essentially, if the Polaris and Poseidon systems were to be anchored at fixed points, they would not represent the threat they pose as mobile systems. The United States had apparently abandoned interest in fixed nuclear installations on the sea bottom, and there is evidence to indicate that the decision to develop post-Polaris deterrent systems rather than fixed nuclear installations had been reached long before the denuclearization of the seabed was considered on the international disarmament agenda.[88] This does not mean, however, that the Navy was not using the sea bottom. In testimony before Representative Dante Fascell's Subcommittee on International Organizations and Movements, Dr. Robert Frosch, Assistant Secretary of the Navy for Research and Development, was asked to describe some of the Navy undertakings which might be involved in the Malta proposal. Dr. Frosch answered "that the Navy has used the sea bottom for many purposes for many years, and it is incorrect to assume that we are not using the sea bottom. Any attempt to deal in a radical legal way with the sea bottom would interfere with some national security enterprise." [89]

Consequently, the factor of national security and the Navy's demands were focal points in formulating the U.S. draft treaty and the final outcome. The technological gap between the United States and the Soviet Union, the high costs of developing underwater systems, and the political developments on the international scene vis-à-vis mainland China were among the other factors shaping the U.S. and U.S.S.R. positions.

Although the time lag between Soviet and U.S. marine capabilities has been considerable in recent years, the gap has been closing at a fast rate. In his annual report on the U.S. military posture, Melvin Laird, Secretary of Defense, indicated that the overall (land, sea, and air) technological challenge from the Soviet Union was so strong as to obliterate any U.S. technology lead over the Soviet Union by the mid-to-late 1970s.[90]

SCIENTISTS IN THE DIPLOMATIC PROCESS

In the section on seabed resources it was indicated how progress in obtaining scientific data and the increasing knowledge of the marine environment produced technological developments that pushed man into progressively deeper waters offshore. Scientific manpower has also been essential in formulating U.S. positions on issues of ocean policy. A number of scientists have participated in advising both the legislative and executive branches of Government. Scientists from academic, industrial, and Government institutions contributed to the formulation of U.S. policy on the seabed. Some have participated in the actual deliberations and drafting of resolutions such as the Draft U.N. Convention on the International Seabed Area.

Role of the Marine Council Staff

Prior to 1966, the Federal effort in marine affairs was distributed among more than 20 agencies, and was for the most part uncoordinated. In the legislative branch, the Library of Congress' Congressional Research Service (then the Legislative Reference Service) established in 1964 the Science Policy Research Division, with Dr. Edward Wenk, Jr., as Chief. Dr. Wenk provided considerable groundwork toward the passage of the Marine Resources Act of 1966, and later became Executive Secretary of the Marine Council, created by the act, under the chairmanship of the Vice President.

Although world attention was focused on the seabed resources following the Malta proposal in 1967, in the United States the Marine Council staff had already been active in laying the groundwork for U.S. policy on this issue. Section 6 of the Marine Resources Act assigned to the Council an explicit responsibility to coordinate a program for international cooperation. Soon after its activation in August 1966, the Council staff guided a series of studies and actions to take into account universally agreed upon goals to which the oceans could contribute, such as to remedy the disparity between world population and food supply. Inquiries were also begun as to threats to world order arising out of conflicts over the extraction of marine resources, and ways and means by which the common interest of all nations in gaining greater knowledge about the marine environment could be served by intergovernmental cooperative programs of ocean research.

By late fall of 1966, the Council staff, working with representatives of the State Department, helped draft a U.S. initiative at the 1966 U.N. General Assembly, calling for an examination of international marine-science activities. By December of the same year, the Council staff understood from U.N. discussions in New York that there was likely to be interest, particularly among the developing nations, in clarifying uncertainties over ocean boundaries through the medium of a new continental shelf convention.[91]

Again on the initiative of the Council staff, and after prior exchanges with State Department staff as to agenda, the Vice President, as Chairman of the Marine Council, met on February 10, 1967, with Deputy Under Secretary of State Foy D. Kohler concerning these issues, with the result that Dean Rusk, Secretary of State, appointed an Ad Hoc Committee for International Policy in the Marine Environment to serve the interests of both the Marine Council and the Department of State. Soon after its formation, this Committee began substantive inquiry into legal regime questions—building on a concept of "revenue belts" or "buffer zones" that had been informally proposed by representatives from the Department of State and the Council.

By that time also, the Council had begun to implement Section 4(a), paragraph 5 of the Marine Resources Act to undertake a comprehensive study of the legal problems arising out of the management, use, development, recovery, and control of the resources of the marine environment. Four contract studies were accordingly undertaken to provide in-house policy guidance. The Vice President requested that the Department of State provide guidelines for these studies, and the Secretary of State appointed an interagency advisory committee chaired by the legal adviser of the Department of State, Leonard Meeker.

When the Committee for International Policy in the Marine Environment met for the first time in April 1967, all of the in-house instruments for the study of the legal regime for the seabed had been created, and some of the directions and alternatives laid out for study—"in turn all goaded by an activist style of the Council itself." [92]

During the early months of the Council, studies were also begun on the International Decade of Ocean Exploration, and seabed disarmament. By late summer of 1967, concepts were beginning to emerge regarding the legal regime, the Decade, and the disarmament issue. These concepts emerged along with conflicts among different Federal agencies as they generated their own independent positions regarding each of the three issues.

Role of Scientists in Other Agencies

For the purpose of this study, inquiries were addressed to several Federal agencies concerning their utilization of scientists in their international policy making. In arms-control negotiations, the Arms Control and Disarmament Agency (ACDA) found the issues complex, involving aspects of legal, scientific, economic, military, and political disciplines. ACDA is organized along these disciplinary lines, and the corresponding bureaus of the agency contributed to the evolutionary formulation of the U.S. position through comprehensive studies and analyses, utilizing internal staff members and cooperating with their counterparts in other agencies. They made extensive use of scientific capabilities of the U.S. Navy, including contributions from their Chief Scientist and the Assistant Secretary of the Navy for Research and Development, Dr. Robert Frosch.

The U.S. Geological Survey participated extensively in the activities at the United Nations, and several Survey geologists have contributed to the work of the Secretariat as well as to that of the U.S. Government. With respect to the United Nations, Frank H. Wang, a geologist with the Survey's Office of Marine Geology, has been loaned to the Resource and Transport Division of the United Nations Secretariat for several periods, beginning in late 1967 and continuing to the present, to prepare a background report on mineral resources of the sea.[93]

In the late spring of 1968, David Popper, then Deputy Assistant Secretary of State for International Organizations, asked the Geological Survey to represent the United States at the Economic and Technical Subcommittee of the newly formed Ad Hoc Committee on the Peaceful Uses of the Seabed Beyond the Limits of National Jurisdiction. The Director of the U.S. Geological Survey, Dr. William Pecora (now Under Secretary of the Interior), was able to attend some of the June meetings, while Vincent E. McKelvey, his alternate and successor, attended the remainder. Gilbert Corwin of the Survey also attended this session as an adviser.

While Dr. Pecora continued to be listed as the U.S. representative to the Economic and Technical Subcommittee during the following year (hoping that by so doing he would encourage other delegations to send high-level scientists), he was unable to attend subsequent meetings of the Subcommittee. Vincent McKelvey, therefore, represented the United States on the Economic and Technical Subcommittee at the second meeting of the Ad Hoc Committee, and has continued to do so at the meetings of the permanent Committee after its establishment by the General Assembly in 1968. Joshua I. Tracey of the Survey assisted McKelvey during the March and August meetings of the Committee in 1969, and Wang was also on the delegation for the August 1969 meeting.

McKelvey was also a member of the five-man drafting committee brought together by John R. Stevenson, Legal Adviser of the Department of State, in June 1970, to prepare the draft treaty implementing the U.S. ocean policy announced by President Nixon on May 23, 1970. McKelvey is the only scientist on that committee, and has been primarily responsible for input not only on the geological aspects of the problem, but also for the economic and technical aspects of seabed exploration and exploitation.

During the first two years of the U.N. Seabed Committee's work, the principal effort of the U.S. Geological Survey was directed toward developing information that would assist other delegations, and the Committee as a whole, in understanding the problems of the seabed. As part of that effort, McKelvey and Wang prepared a set of maps showing the distribution of potential subsea mineral resources, the first edition of which was distributed to the Committee in August 1969.[94] Another contribution to this effort was the Symposium on Mineral Resources of the World Ocean, held at Newport, Rhode Island, in 1968, under the joint sponsorship of the U.S. Geological Survey, the University of Rhode Island, and the U.S. Navy.[95]

Besides the staffs of the Marine Council and the Geological Survey, the Department of State had available the expertise of its own geographer, Dr. Robert Hodgson, who was intimately associated with seabed activities on an official basis for over a decade, and its Bureau of International Scientific and Technological Affairs, under the direction of Herman Pollack. Together with the National Science Foundation, these sources have been represented on the U.S. Government's Law-of-the-Sea Task Force since its creation in 1970. The State Department drew further on the following agencies and scientists, in varying degrees, in the formulation of the administration's ocean policy: Dr. Bruce C. Heezen, Columbia University; Hollis Hedberg, Princeton University; Howard R. Gould, Esso Corporation, Houston; K. O. Emery, Woods Hole Oceanographic Institute, Massachusetts; John Byrne, Oregon State University; John Knauss, University of Rhode Island; and the National Oceanography Association.

SCIENTIFIC ADVICE, POLICY, AND DIPLOMACY

It is particularly true of a democratic society that conflicts arise among parties engaging in the formulation of national and international policy. When the matter at issue involves diplomacy and international negotiation, conflicts are particularly prone to impede the formulation of a generally accepted position. National honor, national security, sovereignty, and territorial claims all combine to intensify feelings and delay the building of a consensus.

Despite the initiatives of the Marine Council staff and the participation of numerous scientists and scientific institutions toward the formulation of U.S. seabed policy, the evolution of this policy was relatively slow. As late as July 1969, almost two years after the Malta proposal, the Department of State had not yet formulated a policy, or was not ready to divulge its position if it had one. Testifying before Senator Pell's Subcommittee on Ocean Space, the Honorable U. Alexis Johnson, Under Secretary of State for Political Affairs, was asked whether the issue of the outer continental shelf boundaries was a question of language or modality.

. . . Frankly, Mr. Chairman [answered Mr. Johnson], the question of the boundaries, the question of the international regime, are questions the answers to which are not yet clear to me, nor am I clear if I may say, both personally and officially, as to where the U.S. interests lie best in this. . . .[96]

Senator Pell termed this a "no-policy policy" in the exchange that ensued:

. . . In closing I would just make the point that I appreciate your frankness and cooperation in coming here today, and I hope you will push ahead with the policy paper for the United Nations meeting.

At the same time, I must stick to my guns, when it comes to the questions of the continental shelf and the moratorium on claims and say that we have a "no-policy" policy, but I am glad to know that you are pressing ahead to change that to a more specific statement of policy. If you think I have overstated the situation, please tell me.

Mr. JOHNSON. No, frankly. I feel we have taken more of a leadership role in this matter than you apparently feel, but nevertheless, I respect your point of view.

Senator PELL. You mean a leadership role for going ahead or a leadership role for going backwards? By this I mean a leadership role for establishing a regime or a leadership role in preventing the establishment of a regime.

Mr. JOHNSON. I would say a leadership role in keeping our options open until we decide where our national interests lie best and where international agreement may be reached.

Senator PELL. Right. Well, I do not want to be rude in any way, but basically, to keep options open, means to my mind to have a "no-policy" policy.

Mr. JOHNSON. That is correct. We are keeping options open for that purpose until we decide what our policy should be on this.

Senator PELL. I agree that this is probably a question of semantics and what the executive branch would call keeping options open, from where I sit and the work that I have been doing on this for the last several years, I would say that it is a "no-policy" policy. I know we are both doing the best that we can to try to arrive at a state of affairs of advantage not only to the United States, but to the world as a whole.

Mr. JOHNSON. Yes.[97]

A policy statement enunciated by the President is transmitted by directive to the departments concerned for implementation. The departments evaluate it relative to their statutory responsibilities, policies, and practices, then try to relate it to the overall national and international perspective, at the same time accommodating their own interests.

Congress, meanwhile, provides a forum where all sectors and individuals are afforded a chance to air their views on the subject. In the case of oceanography, Congress has had the initiative for more than a decade, and its efforts culminated in the passage of the Marine Resources Act of 1966, despite some opposition by the executive branch. However, not all of the views expressed at hearings are thoroughly studied, or influence final national policy in any real way unless such views are vigorously pursued and advocated by special interest groups.

In ocean affairs, the "ocean" industry in general lacks a unified front, or a spokesman or representative in Washington capable of presenting the industry's point of view. It remains as uncoordinated as were the Federal agencies prior to the establishment of the National Council on Marine Resources and Engineering Development, and the National Oceanic and Atmospheric Administration (NOAA). Although a NOAA had been recommended by the Commission on Marine Science, Engineering, and Resources, the NOAA that came into existence in October 1970 fell short of the Commission's recommendations, leaving many ocean activities scattered among other Federal agencies.

The machinery of most governments does not provide adequately for coordination between individuals qualified to judge in the real world of politics and people, and those qualified to judge in the real world of technical facts. Often, one world seems to be completely oblivious of the existence of the other. It has been demonstrated, however, that scientists can work very effectively in formulating policy and participating in the diplomatic process. Modern diplomats are becoming increasingly aware of the effect of science and technology on shaping their daily endeavors. Although most nations have come to recognize the importance of scientists in conducting their international affairs, no country seems to have included a scientist as part of its diplomatic staff at the United Nations headquarters in New York.

10

Overview

The earth is essentially a water planet—one large ocean interspersed with continental land masses. The global ocean is a common link among these land masses, shared by the nations touching this ocean space. Despite its inherent international characteristics, ocean space has been zoned off, and national jurisdictions and boundaries have been established by the coastal states.

Progress in marine technology and the widening horizons of scientific inquiry have enlarged the sphere of man's knowledge and revealed the presence of natural resources in sea water, on the ocean floor, and in the underlying layers. Peace and equity require internationally acceptable boundaries and definitions of territorial limits, fishing zones, the high seas, the continental shelf, and the sea floor beyond the limits of national jurisdiction. It has become necessary to survey the ocean space, to collect the scientific data on which these definitions should be based, and to inventory the known and potential resources of the seabed.

The crust of the earth as a whole is composed of continental platforms and ocean basins. Geologically, the continental land masses extend beyond the shoreline. A relatively narrow strip, the "continental margin," of each platform, is under water, belonging geologically to the adjacent continent and not to the ocean basin. The continental margin has three components: the shelf, the slope, and the rise. The width of the shelf varies throughout the world, but an average water depth of 100 fathoms (600 feet) has been adopted as conveniently marking the legal, rather than the geological, width of the continental shelf.

Unilateral actions have been taken by coastal nations to assert jurisdiction, establish territorial boundaries, and claim exploration and exploitation rights in offshore areas. In the United States, these activities began with the Truman Proclamation of 1945, which claimed the natural resources of the seabed of the continental shelf as appertaining to the United States and subject to its jurisdiction and control. In 1953, the Submerged Lands Act set the seaward limit of state boundaries as three miles, but did not define inland waters or continental-shelf lands beyond the three-mile limit. The Outer Continental Shelf Lands Act of 1953 claimed for the United States

rights of jurisdiction, control, and power of disposition of the natural resources of the continental shelf, but left the seaward limits of the shelf undefined, and preserved the character of the overlying waters as high seas.

The Geneva Conventions of 1958 sought to resolve several problems
pertaining to the seabed and the overlying waters. The conventions established criteria for measuring the territorial sea and the contiguous zone, but
left undefined the outer limits of the continental shelf. The Convention on
the Continental Shelf aggravated the problem further by establishing the
200-meter depth as the recommended limit, which could be expanded
beyond that depth where the depth of the superjacent waters admits of the
exploitation of the natural resources of that area. In other words, it proposed that the boundaries of the continental shelf of a coastal state would
be determined by the technological capabilities of that state to exploit the
resources in deeper waters.

But what are these resources, and what exactly is their present and
prospective value? What is the present state of offshore technology, and
what lies ahead for future exploration and exploitation of the seabed?

Ocean resources are classified broadly as living and nonliving. The
living resources include the living organisms of the marine environment
sought for products such as food, food derivatives, and pharmaceuticals.
The nonliving resources provide such varied opportunities for use as the
production of potable water from the sea, the salts and other minerals contained in the water, the minerals on and under the ocean floor, and such
related activities as shipping and aquatic recreation.

Although this study encompasses the living resources of the sea, it
focuses on the seabed and the resources contained in, on, and under it. The
seabed contains a variety of mineral resources, including beach sands and
gravel, heavy minerals associated with beach deposits, surface deposits of
manganese and phosphorite, and subsurface petroleum resources.

Building materials, the most extensively mined commodity throughout
the world, are mined mostly at or near the beaches. Current production in
the United States alone exceeds 50 million cubic yards of sand and gravel,
and 20 million tons of oyster shells annually. Associated with beach sands
are such heavy minerals as gold, tin, platinum, diamonds, titanium, tungsten, iron, chromite, and zircon. Surface deposits of phosphorite and
manganese nodules blanket the ocean floor. The continental shelves of the
world contain an estimated 300 billion tons of phosphorite; if 10 percent of
this amount is economic to mine, the 30 billion tons of reserves of sea-floor
phosphorite (worth something like $300 billion) would last 1,000 years.

Equally extensive on the ocean floor are nodules containing manga-
nese and iron oxide, cobalt, nickel, and copper. The floor of the Pacific
Ocean alone contains some 90 billion to 1,600 billion tons of nodules.
Although submarine manganese ore is lower in grade than manganese

mined on land, its mining may become attractive for its combination of useful elements. Gross value of the constituent metals in these nodules may exceed $115 per ton. Although under the proper circumstances the potential of these nodules might be promising, present factors of supply, demand, and pricing suggest that large-scale economic exploitation of phosphorite and manganese nodules is not likely in the immediate future.

Recently discovered deposits in the Red Sea indicate that volcanic action in areas of rifts in the crust of the earth may have created economically profitable opportunities. Sediments sampled in the Red Sea contain appreciable amounts of zinc, copper, lead, silver, and gold which at current smelter prices would be worth about $2.5 billion.

By far the most important of all marine resources is petroleum. More than 85 countries are engaged in offshore activities; discoveries have been reported from the shelves of North and South America, Australia, Japan, the Mediterranean countries, the Red Sea, the Arabian Gulf, the Union of Soviet Socialist Republics and, most recently, in the North Sea and the South China Sea. Thirty-two of these countries are already producing petroleum from their continental shelves, which accounts for 16 percent of the world's oil and 6 percent of the world's natural gas; by 1980 this percentage is expected to double or quadruple.

Proved petroleum reserves in the "free world" are estimated to exceed 500 billion barrels of oil and nearly 1.5 million billion (quadrillion) cubic feet of gas. Ultimate world potential of all *offshore* petroleum resources approaches 1,600 billion barrels. In comparison, ultimate world potential of comparable resources *on land* is estimated at 4,000 billion barrels.

Given the abundant resources of the seabed and the challenges they present, especially to the dynamic petroleum industry, what are their implications for national policy and international diplomacy? In mining operations, a foremost consideration is the necessity for maintaining an approximate balance of supply against demand. A prime economic characteristic of all minerals, except those that are scarce and precious, is their price sensitivity. The risk, the high capital investment, the unknowns, and the lack of experience in the marine environment are major deterrents which will keep the small entrepreneur from venturing into the deep-sea operations. Another deterrent, nontechnical and noneconomic, is the legal question of exclusive right of exploitation, or security of tenure of operation. Undoubtedly, deep-sea mineral deposits are substantial and represent a great potential resource. Once a substantially rich deposit is found, the technology to exploit it will be readily developed. The outlook is one of cautious optimism, but a legal regime needs to be established and international agreements effected before the mining industry will venture into the ocean deeps.

For offshore petroleum, the capabilities that are not available now are certainly a short distance away. But does technology justify expansion? As matters stand now, offshore operations seem likely to continue at an ever-increasing pace. While subject to some degree of technological control, hazards of offshore operations are inevitable, and damage to the environment may be long-lasting or irreversible. Numerous other tenants besides the oil industry use coastal waters. The ocean has become the focus of man's attention and hope, not merely for its mineral and petroleum resources, but as a source of food, a possible future habitat, and a major source of the earth's weather systems and their life-giving processes. All users of the continental shelf have to recognize the variety of their activities, and strive to achieve compatability. It may be in the ultimate interest of all mankind to develop the land areas and explore their subsurface thoroughly, leaving the ocean as clean as possible for as long as possible.

This go-slow policy is particularly crucial for the continental shelf in view of the fact that technological development is progressing at a rate that has already rendered obsolete the definition of jurisdictional limits, legal or otherwise. While this development will probably be limited, for some time to come, to the continental shelf areas, and progress into the deep sea is not alarmingly imminent, the confusion created by the Geneva Conventions, particularly the exploitability clause, might well be eliminated. Definitive political boundaries need to be established for the seaward limit of national jurisdictions. Beyond this limit, the deep-sea areas would then become the common domain of the community of nations.

The rapid advances in the acquisition of scientific data about the ocean domain, and the spectacular development of technological capabilities to exploit it commercially and militarily have directed attention to the potential of ocean resources. As nations have moved toward a policy of leaving ocean space free from national domination, the aspiration has been repeatedly voiced of exploring and exploiting ocean resources for the benefit of all mankind, rather than to benefit the handful of technologically advanced nations.

The United Nations is the obvious forum to reconcile issues over these resources. A specific plan was offered to the United Nations by the delegation of Malta, which called for a declaration and treaty concerning the reservation exclusively for peaceful purposes of the seabed and of the ocean floor underlying the seas beyond the limits of present national jurisdiction, and the use of their resources in the interest of mankind. An ad hoc committee to study this proposal was formed in 1967, which became in 1968 the Committee on the Peaceful Uses of the Sea-Bed and the Ocean Floor beyond the Limits of National Jurisdiction.

By then the United States Congress had passed the Marine Resources and Engineering Development Act of 1966 and established the National Council on Marine Resources and Engineering Development; the U.S. Government had begun to coordinate its ocean affairs and formulate policy for participation in international activities. Besides the Marine Council, the policy apparatus included committees of Congress, the Committee on International Policy in the Marine Environment, and the present Interagency Law-of-the-Sea Task Force. Outside the Federal structure, the United States sought advice from the National Academy of Sciences and the National Academy of Engineering.

Congressional reaction in the 90th Congress to the Malta proposal took the form of numerous bills and resolutions, some in support of and others in opposition to the proposal. Hearings were conducted in the 90th, 91st, and 92nd Congresses, and new subcommittees were established, particularly in the Senate. The Subcommittee on Outer Continental Shelf of the Senate Committee on Interior and Insular Affairs held extensive hearings throughout the 91st Congress and issued a report based on these hearings. The subcommittee took the position that the Geneva Convention on the Continental Shelf was validly operative, and saw no need to convene another Law of the Sea Conference. It also concluded that the geological interpretation of the continental margin made that portion of the seabed the property of the United States. It endorsed the exploitability clause in the convention, which made expansion of the limits of the shelf dependent on the technological capability of a state to exploit in deeper waters. It shared with the President the expressed desire that ocean resources *beyond* the continental margin be used rationally and equitably for the benefit of mankind, conditional on measures to protect investors exercising high-seas rights to explore and exploit the wealth of the deep seabed.

The executive branch leaned toward international cooperation and the proper utilization of the diplomatic process. U.S. delegates to the United Nations General Assembly took the initiative in introducing several draft resolutions toward international cooperation in research, the exploitation of the seabed, and the limitations of military uses of the sea floor. These efforts culminated in the signing of the Seabed Disarmament Treaty on February 12, 1971, banning the emplacement of nuclear weapons on the ocean floor, and paving the way for wider measures toward disarmament.

As to an international seabed regime, President Nixon proposed on May 23, 1970, that all nations adopt as soon as possible a treaty renouncing all national claims over the natural resources of the seabed beyond the point where the high seas reach a depth of 200 meters, and agree to regard these resources as the common heritage of mankind. The regime proposed for the exploitation of seabed resources would provide for the collection of substantial mineral royalties to be used for international community purposes, particularly for economic assistance to developing countries. It would also establish rules and regulations for protecting the ocean environment, and a mechanism for the settlement of disputes, in the form of an International

Seabed Resource Authority. In the meantime, an interim policy was pro-posed for all nations to join the United States to insure that all permits for exploration and exploitation of the seabed beyond the 200-meter limit be issued subject to an international authority.

During the 25th session of the U.N. General Assembly, these prin-ciples were considered, and on December 18, 1970, two resolutions were passed: One established a timetable and called for convening in 1973 of a new conference on the law of the sea; the other promulgated a set of prin-ciples in a declaration of ground rules for ocean resources management and scientific research.

In its diplomatic participation, the United States developed policy contingent on developments in science and technology. Since World War II, the outlook toward the use of the oceans for military purposes has been gaining progressively larger dimensions. Military strategy has evolved along lines drawn by developments in technology and by policy goals for internal security and global policies. Communist China's progress in diplomatic, economic, and nuclear status may also explain the urgency and pressure to resolve issues of territorial limits, continental shelf boundaries, and seabed resources, as well as the banning of nuclear weapons from the ocean floor.

In formulating policy, the United States has had the benefit of con-siderable scientific guidance. A number of scientists have participated in advising both the legislative and the executive branches of Government. Scientists from academic, industrial, and Government institutions were in-strumental in the formulation of U.S. policy on the seabed. Some scientists have participated in the actual deliberation and drafting of resolutions such as the Draft U.N. Convention on the International Seabed Area.

Despite the initiatives of the Marine Council staff and the increased participation of scientists in the formulation of U.S. seabed policy, the evolution of this policy has been relatively slow. Undoubtedly the marine scientists and technologists would have preferred a brisker pace than the diplomats were prepared to take. For its part, the Congress was ready to move faster than was the Department of State, although in what direction is still not evident. In the case of oceanography, Congress has had the initia-tive for more than a decade; its efforts culminated in passage of the Marine Resources Act of 1966, despite outright opposition by the executive branch.

It has been demonstrated that scientists can work effectively in helping to formulate policy and in participating in the diplomatic process. Modern diplomats are becoming increasingly aware of the effect of science and technology in shaping their endeavors. The diplomatic process is in some ways inherently ambiguous and indirect. Traditional diplomatic ambiguity is often difficult to reconcile with scientific precision and explicitness, and few persons can combine the subtleties and intuitive approach of the diplomat with the straightforward factual approach of the scientist to perform adequately across both fields. Nevertheless, the number of those who can—the new breed of scientist-diplomat, or policy-making scientist—is rising rapidly. It is to the advantage of a nation to capitalize on the skills of such individuals in the pursuit of the national interest, for they may represent mankind's hope for the effective conduct of decision making in a world society of nations increasingly interdependent and influenced by scientific discovery and technological change.

References

1. *Congressional Record,* daily ed., April 3, 1968, p. S3818.
2. CEP Doc. No. 2, September 23, 1955, contained in "Santiago Negotiations on Fishery Conservation Problems," U.S. Department of State, Public Service Division, 1955, pp. 30–32.
3. Francis T. Christy, Jr. and Anthony Scott, *The Common Wealth in Ocean Fisheries: Some Problems of Growth and Economic Allocation,* Baltimore, Johns Hopkins Press, 1965, p. 163.
4. Wilbert M. Chapman, "A Symposium on National Interests in Coastal Waters," in *The Law of the Sea,* ed. by Lewis M. Alexander, Ohio State University Press, 1967, p. 125.
5. *Outer Continental Shelf Lands Act,* report of the Committee on Interior and Insular Affairs, Senate Report No. 411, 83rd Cong., 1st Sess., 1953, June 15, 1953, p. 4.
6. *Ibid.,* p. 5.
7. League of Nations Documents C.74.M.39.1929.V and 1930.V.
8. "International Law Commission," report, United Nations General Assembly, Official Record, 11th Sess., Supp. No. 9 (A/3159), 1956.
9. *Ibid.,* Article 3.
10. John L. Mero, *The Mineral Resources of the Sea,* New York, Elsevier, 1965, p. 71.
11. Vincent E. McKelvey, "Mineral Potential of the Submerged Parts of the Continents," in *Mineral Resources of the World Ocean,* proceedings of a symposium held at the Naval War College, Newport, R.I., July 11–12, 1968, ed. by Elizabeth Keiffer, University of Rhode Island, Occasional Publications No. 4, 1968, p. 34
12. Mero, *op. cit.,* p. 73.
13. D. D. Kvasov, "Limnological Hypothesis of the Origin of Hot Brines in the Red Sea," *Nature,* 221:850–851, March 1, 1969.
14. J. L. Bischoff and F. T. Manheim, "Economic Potential of the Red Sea Heavy Metal Deposits," in *Hot Brines and Recent Heavy Metal Deposits in the Red Sea: A Geochemical and Geophysical Account,* ed. by E. T. Degens and D. A. Ross, New York, Springer-Verlag, 1969, p. 535.
15. Joseph F. Stevens, "Mining the Alaskan Seas," *Ocean Industry,* November 1970, pp. 47–49.
16. *Commodity Data Summaries,* U.S. Bureau of Mines, January 1971, pp. 10–11.
17. *Oil and Gas Journal,* December 29, 1969, p. 95.
18. E. D. Schneider and G. L. Johnson, "Deep-Ocean Diapir Occurrences," *American Association of Petroleum Geologists Bulletin,* November 1970, pp. 2151, 2169; see also H. K. Wong *et al.,* "Newly Discovered Group of

Diapiric Structures in Western Mediterranean," *American Association of Petroleum Geologists Bulletin,* November 1970, pp. 2200–2204.

19. M. Ewing *et al.,* "Site 2," in *Initial Reports of the Deep Sea Drilling Project,* Vol. I, Lamont-Doherty Geological Observatory Contribution No. 1364, June 1969, pp. 84–111.

20. C. P. Idyll, *The Sea Against Hunger,* New York, Thomas Y. Crowell, 1970, pp. 22–24.

21. Milton S. Sachs, *Desalting Plants, Inventory Report No. 1,* report of Office of Saline Water, U.S. Department of the Interior, Washington, D.C., U.S. Government Printing Office, January 1, 1968.

22. Mero, *op. cit.,* p. 4.

23. Frank Wang, *Mineral Resources of the Sea,* U.N. Department of Economic and Social Affairs, ST/ECA/125, April 1970, p. 4. This is an updated version of an identical report of the Secretary General dated Februray 19, 1968.

24. Mero, *op. cit.,* p. 42.

25. W. F. McIlhenny and D. A. Ballard, "The Sea As a Source of Dissolved Chemicals," in *Symposia on Economic Importance of Chemicals from the Sea,* Washington, Division of Chemical Marketing Economics, American Chemical Society, 1963.

26. Gordon O. Pherson, "Mining Industry's Role in Development of Undersea Mining," in *Exploiting the Ocean,* transactions of the 2nd Annual Marine Technology Conference and Exhibit, June 27–29, 1966, p. 195.

27. Mero, *op. cit.,* p. 257.

28. M. P. Overall, "Mining Phosphorite from the Sea, Part ?: Economics of Mining and Beneficiation," *Ocean Industry,* October 1968, p. 60.

29. Chester O. Ensign, Jr., "Economic Barriers Delay Undersea Mining," *Mining Engineering,* September 1966, p. 60.

30. *Ibid.,* p. 61.

31. Overall, *op. cit.,* p. 61.

32. *Ibid.,* p. 60.

33. This information was revealed in a statement by Hashim Dabbas, Jordanian Under Secretary of the Ministry of Economy, reported in the newspaper *Ar-Ra'y* (in Arabic), Amman, January 6, 1972, p. 2.

34. David B. Brooks, *Low-Grade and Nonconventional Sources of Manganese,* Baltimore, Johns Hopkins Press, 1966.

35. *Ibid.,* p. 103.

36. *Commodity Data Summaries,* January 1971, pp. 42, 88, 102.

37. Mero, *op. cit.,* p. 279.

38. Quoted by *Ocean Science News,* June 12, 1970, from a presentation by LaQue at the Marine Technology Society meeting. Similar calculations were submitted by LaQue at the Conference at Ditchley Park, September 26–29, 1969. See K. R. Simmonds, *The Resources of the Ocean Bed,* 1969, pp. 15–17.

39. Brooks, *op. cit.,* p. 106.

40. *Ibid.,* pp. 106–107.
41. Richard C. Sparling, Norma J. Anderson, and John G. Winger, "Capital Investment of the World Petroleum Industry—1969," The Chase Manhattan Bank, December 1970, pp. 2, 19.
42. John D. Emerson, "The Petroleum Situation in October 1970," Energy Division, The Chase Manhattan Bank, November 30, 1970, pp. 2–3.
43. Eduardo J. Guzman, "Are Sophisticated Exploration Methods the Answer?" in *Exploration and Economics of the Petroleum Industry: New Ideas, New Methods, New Developments,* ed. by Virginia S. Cameron, 1969, pp. 21–22.
44. Lewis G. Weeks, "The Gas, Oil and Sulfur Potentials of the Sea," *Ocean Industry,* June 1968, p. 46; see also Chris Welles, *The Elusive Bonanza: The Story of Oil Shale,* New York, Dutton, 1970.
45. Speech given at the commissioning of the research ship *Oceanographer* at the Washington Navy Yard on July 13, 1966.
46. *The United Nations at Twenty-one,* Senate Committee on Foreign Relations, report by Senator Frank Church, 90th Cong., 1st Sess., 1967, p. 25.
47. *Marine Science and Technology: Survey and Proposals,* U.N. Economic and Social Council, New York, April 24, 1968, pp. 50–59 and Annex X (UNESCO, E/4487).
48. "Malta: Request for the Inclusion of a Supplementary Item in the Agenda of the Twenty-second Session," United Nations, New York, August 18, 1967, U.N. Document A/6695.
49. *Resources of the Sea (Beyond the Continental Shelf),* report of the Secretary General, United Nations, New York, February 21, 1968, U.N. Economic and Social Council Document E/4449; Part One: "Mineral Resources of the Sea Beyond the Continental Shelf," E/4449/Add. 1; Part Two: "Food Resources of the Sea Beyond the Continental Shelf Excluding Fish," E/4449/Add. 2. Part One was prepared jointly by Frank Wang, marine geologist of the U.S. Geological Survey and the United Nations Secretariat. Part Two was prepared by C. P. Idyll of the Institute of Marine Sciences, University of Miami, Florida.
50. *Marine Science and Technology: Survey and Proposals,* United Nations, New York, April 24, 1968, Document UNESCO E/4487.
51. *Study on the Question of Establishing in Due Time Appropriate International Machinery for the Promotion of the Exploration and Exploitation of the Resources of the Seabed and the Ocean Floor Beyond the Limits of National Jurisdiction and the Use of These Resources in the Interests of Mankind,* Secretary General, United Nations, New York, A/AC.138/12.
52. *News Report,* National Academy of Sciences, National Research Council, National Academy of Engineering, Vol. 20, No. 9, November 1970.
53. *The United Nations and the Issue of Deep Ocean Resources: Interim Report Together with Hearings,* House Committee on Foreign Affairs, Subcommittee on International Organizations and Movements of the . . . on H.J. Res. 816 and companion resolutions, H.R. No. 999, 90th Cong., 1st Sess., October 10, 19, 25, and 31, 1967.
54. *Governing the Use of Ocean Space,* Senate Committee on Foreign Relations

hearings on S.J. Res. 111, S. Res. 172, and S. Res. 196, 90th Cong., 1st Sess., November 29, 1967.

55. *The Oceans: A Challenging New Frontier,* report, together with hearings and additional documents and materials, by the House Committee on Foreign Affairs, Subcommittee on International Organizations and Movements, 90th Cong., 2nd Sess., June 12, July 25, 1968.

56. *The United Nations and the Issue of Deep Ocean Resources,* p. 5.

57. *Special Study on United Nations Suboceanic Lands Policy,* Senate Committee on Commerce hearings, 91st Cong., 1st Sess., September 23, 24, October 3, and November 21, 1969.

58. *Governing the Use of Ocean Space,* hearings.

59. *Special Study on United Nations Suboceanic Lands Policy,* hearings.

60. *The United Nations and the Issue of Deep Ocean Resources,* p. 68.

61. *Our Nation and the Sea,* report of the Commission on Marine Science, Engineering, and Resources, Washington, D.C., U.S. Government Printing Office, January 1969.

62. *Governing the Use of Ocean Space,* hearings.

63. *Special Study on United Nations Suboceanic Lands Policy,* hearings.

64. *Outer Continental Shelf,* hearings by the Special Subcommittee on Outer Continental Shelf, Senate Committee on Interior and Insular Affairs, Parts 1–3, 91st Cong., 1st and 2nd Sess., 1969, 1970. (Part 3 contains the hearings continued by the Subcommittee on Minerals, Materials, and Fuels on the same subject.)

65. *Ibid.,* p. 380.

66. *Ibid.,* p. 395.

67. *Ibid.,* p. 396.

68. Claiborne Pell, "Barring of Nuclear Weapons from the Seabed and Ocean Floor," statement of the Hon. Claiborne Pell on the floor of the Senate, *Congressional Record,* September 16, 1970, p. S15616.

69. *Outer Continental Shelf,* report by the Special Subcommittee on Outer Continental Shelf, Senate Committee on Interior and Insular Affairs, 91st Cong., 2nd Sess., December 21, 1970.

70. *Ibid.,* p. 3.

71. *Ibid.,* p. 5

72. *Ibid.,* p. 16.

73. *Ibid.,* pp. 16–17.

74. *Ibid.,* p. 30.

75. *Ibid.,* p. 32.

76. *Ibid.,* p. 33.

77. *Op. cit.*

78. *Effective Use of the Sea,* report of the Panel on Oceanography, U.S. President's Science Advisory Committee, Washington, D.C., U.S. Government Printing Office, June 1966, p. 30.

79. Robert A. Frosch, "Exploiting Marine Mineral Resources: Problems of National Security and Jurisdiction," speech delivered at the Naval War College Conference on Mineral Resources of the World Ocean, July 12, 1968, *Vital Speeches of the Day,* November 15, 1968, p. 71.

80. Presidential announcement on U.S. ocean policy, May 23, 1970 (see Appendix 16).

81. *Freedom of Scientific Research and Exploration of the Sea and the Seabed,* news release by the Committee on Oceanography, National Academy of Sciences, National Research Council, June 11, 1970.

82. Ambassador Christopher H. Phillips, "Statement to the U.N. Committee on the Peaceful Uses of the Seabed and the Ocean Floor beyond the Limits of National Jurisdiction," Palais des Nations, Geneva, August 28, 1970.

83. *The Law of the Sea Crisis: A Staff Report on the United Nations Seabed Committee, the Outer Continental Shelf and Marine Development,* Senate Committee on Interior and Insular Affairs, 92nd Cong., 1st Sess., committee print, Washington, D.C., U.S. Government Printing Office, December 1971 (1972), p. 10.

84. *Ibid.,* p. 5.

85. George E. Lowe, "The Only Option?" *U.S. Naval Institute Proceedings,* April 1971, p. 23.

86. Gordon J. F. MacDonald, "An American Strategy for the Oceans," in *Uses of the Seas,* American Assembly, New York, Columbia University Press, 1968, pp. 183–184.

87. *Activities of Nations in Ocean Space,* hearings before the Subcommittee on Ocean Space, Senate Committee on Foreign Relations, on S. Res. 33, 91st Cong., 1st Sess., July 24, 25, 28, and 30, 1969, pp. 45–46.

88. Jozef Goldblat, "The Militarization of the Deep Ocean: The Sea-Bed Treaty," in *SIPRI Yearbook of World Armaments and Disarmament, 1969/1970,* Stockholm, Stockholm International Peace Research Institute 1970, p. 176.

89. *The United Nations and the Issue of Deep Ocean Resources,* p. 192.

90. *Ocean Science News,* March 12, 1971, p. 3.

91. This information about the in-house activities of the Council staff was supplied through personal communication by Dr. Edward Wenk, Jr., former Executive Secretary of the Marine Council. Further details appear in Dr. Wenk's book *The Politics of the Ocean,* University of Washington Press, 1972.

92. *Ibid.*

93. Wang, *Mineral Resources of the Sea.*

94. Vincent E. McKelvey and Frank F. H. Wang, *Preliminary Maps, World Subsea Mineral Resources: Miscellaneous Geological Investigations, Map I–632,* U.S. Geological Survey, Department of the Interior, 1969.

95. McKelvey, *op. cit.*

96. *Governing the Use of Ocean Space,* hearings, pp. 221–222.

97. *Ibid.,* p. 232.

Appendix 1

Breadth of Territorial Seas and Other Jurisdictions Claimed by Most Countries

Country	Territorial Sea (Miles)	Fishing Limit (Miles)	Other Claims
Albania	12	12	
Algeria	12	12	
Argentina	200	200	Sovereignty is claimed over a 200-mile maritime zone, but the law specifically provides that freedom of navigation of ships and aircraft in the zone is unaffected. Continental shelf — including sovereignty over superjacent waters.
Australia	3	12	
Bahrain	3	—	Seabed and subsoil extending seaward to boundaries to be determined.
Barbados	3	—	Follow UK Continental Shelf Convention.
Belgium	3	12	
Brazil	200	12	
Bulgaria	12	12	
Burma	12	12	
Cambodia	12	12	Continental shelf claimed to 50 miles, including sovereignty over superjacent waters.
Cameroon	18	18	
Canada	12	12	
Ceylon	12	6	Claims right to establish conservation zones within 100 nautical miles of the territorial sea.
Chile	50²	200	
China, Peoples Republic of	12	12	
China, Republic of	3	3	Claim made to shallow-water areas and resources in East China and Yellow Seas.
Colombia	12	12	
Congo (Brazzaville)	12	3 to 15	
Congo (Kinshasa. See Zaire)			
Costa Rica	3	—	"Specialized competence" over living resources to 200 miles.

Country	Territorial Sea (Miles)	Fishing Limit (Miles)	Other Claims
Cuba	3	3	
Cyprus	12	12	
Dahomey	12	12	100-mile mineral exploitation limit.
Denmark	3	12[1]	
Greenland	—	12	
Faroe Islands	—	12	
Dominican Republic	6	12	Contiguous zone 6 miles beyond territorial sea for protection of health, fiscal, customs matters, and the conservation of fisheries and other natural resources of the sea.
Ecuador	200	200	
Egypt	12	12	
El Salvador	200	200	
Ethiopia	12	12	
Fiji	3	—	Party to the Territorial Sea Convention (Mar. 1971).
Finland	4	4	
France	3	12	
Gabon	25	25	
Gambia	50	18	
Germany, East	3	3	
Germany, Federal Republic of	3	12[1]	
Ghana	12	12	Undefined protective areas may be proclaimed seaward of territorial sea, and up to 100 miles seaward of territorial sea may be proclaimed fishing conservation zone.
Greece	6	6	
Guatemala	12	12	
Guinea	130	12	
Guinea, Equatorial	12	—	
Guyana	3	3	
Haiti	6	6	
Honduras	12	12	
Iceland	4	12	
India	12	12	Plus right to establish 100 miles conservation zone.
Indonesia	12	12	Archipelago-concept baselines.
Iran	12	12	
Iraq	12	12	
Ireland	3	12[1]	
Israel	6	6	
Italy	6	12[1]	
Ivory Coast	6	12	
Jamaica	12	12	
Japan	3	3	
Jordan	3	3	

Country	Territorial Sea (Miles)	Fishing Limit (Miles)	Other Claims
Kenya	12	12	
Korea, North	12	—	
Korea	20 to 200	20 to 200	Continental shelf—including sovereignty over superjacent waters.
Kuwait	12	12	
Lebanon	—	6	
Liberia	12	12	
Libya	12	12	
Malagasy Republic	12	12	
Malaysia	12	12	
Maldive Islands	3 to 55	100 to 150	
Malta	6	12	
Mauritania	12	12	
Mauritius	12	—	
Mexico	12	12	
Monaco	3	12	
Morocco	12	12	Exception—6 mile fishing zone for Strait of Gibraltar.
Netherlands	3	12[1]	
New Zealand	3	12	
Nicaragua	3	200	Continental shelf—including sovereignty over superjacent waters.
Nigeria	30	30	
Norway	4	12	
Oman	3	—	
Pakistan	12	12	Plus right to establish 100-mile conservation zones.
Panama	200	200	Continental shelf—including sovereignty over superjacent waters.
Peru	200	200	
Philippines	—	—	Archipelago-concept baselines. Waters between these baselines and the limits described in the Treaty of Paris, Dec. 10, 1898; the United States-Spain Treaty of Nov. 7, 1900; and United States-United Kingdom Treaty of Jan. 2, 1930, are claimed as territorial sea.
Poland	3	12	
Portugal	—	12[1]	
Qatar	3	—	
Romania	12	12	
Saudi Arabia	12	12	
Senegal	12	18	Fishing zone beyond 12 miles does not apply to those nations which are party to the 1958 Geneva Convention on the Territorial Sea and the Contiguous Zone.
Sierra Leone	200	12	
Singapore	3	3	

Country	Territorial Sea (Miles)	Fishing Limit (Miles)	Other Claims
Somali Republic	12	12	
South Africa	6	12	
Spain	6	12[1]	
Sudan	12	12	
Sweden	4	12[1]	
Syria	12	12	Contiguous zone—an additional 6-mile area to control security, customs, hygiene, and financial matters.
Tanzania	12	12	
Thailand	12	12	
Togo	12	12	
Trinidad and Tobago	12	12	
Tunisia	6	12	Fisheries zone follows the 50-meter isobath at specified areas of the coast (maximum 65 miles).
Turkey	6	12	Territorial sea 12 miles in Black Sea.
Ukrainian S.S.R.	12	12	
Union of Arab Emirates:			
Abu Dhabi	3	—	
Ajman	3	—	
Dubai	3	—	
Fujairah	3	—	
Ras al Khaimah	3	—	
Sharjah	12	—	
Umm al Qaiwain	3	—	
U.S.S.R.	12	12	
United Kingdom	3	12	
Overseas areas	3	3	
United States of America	3	12	
Uruguay	200	12	Sovereignty is claimed over a 200-mile maritime zone, but law specifically provides that the freedom of navigation of ships and aircraft beyond 12 miles is unaffected by the claim.
Venezuela	12	12	
Vietnam, North	12	20[2]	
Vietnam, Republic of	3	20[2]	
Yemen Arab Republic	12	12	
Yemen, Peoples Democratic Republic of	12	12	Continental shelf to 300 meters.
Yugoslavia	10	10	
Zaire	3	3	

[1] Parties to the European Fisheries Convention, which provides for the right to establish a 3-mile exclusive fishing zone seaward of 3-mile territorial sea plus additional 6-mile fishing zone restricted to the Convention nations.

[2] Kilometers (1 km = 0.62 mi).

Source: Information available from the U.S. Department of State, including *International Boundary Study, Series A, Limits in the Seas; National Claims to Maritime Jurisdictions,* Bureau of Intelligence and Research, No. 36, Jan. 3, 1972.

Appendix 2

A Proclamation (No. 2667):
Policy of the United States with Respect to the
Natural Resources of the Subsoil and Seabed
of the Continental Shelf

(By the President of the United States of America)

Whereas the Government of the United States of America, aware of the long-range world-wide need for new sources of petroleum and other minerals, holds the view that efforts to discover and make available new supplies of these resources should be encouraged; and

Whereas its competent experts are of the opinion that such resources underlie many parts of the continental shelf off the coasts of the United States of America, and that with modern technological progress their utilization is already practicable or will become so at an early date; and

Whereas recognized jurisdiction over these resources is required in the interest of their conservation and prudent utilization when and as development is undertaken; and

Whereas it is the view of the Government of the United States that the exercise of jurisdiction over the natural resources of the subsoil and sea bed of the continental shelf by the contiguous nation is reasonable and just, since the effectiveness of measures to utilize or conserve these resources would be contingent upon cooperation and protection from the shore, since the continental shelf may be regarded as an extension of the land-mass of the coastal nation and thus naturally appurtenant to it, since these resources frequently form a seaward extension of a pool or deposit lying within the territory, and since self-protection compels the coastal nation to keep close watch over activities off its shores which are of the nature necessary for utilization of these resources;

Now, Therefore, I, Harry S. Truman, President of the United States of America, do hereby proclaim the following policy of the United States of

America with respect to the natural resources of the subsoil and sea bed of the continental shelf.

Having concern for the urgency of conserving and prudently utilizing its natural resources, the Government of the United States regards the natural resources of the subsoil and sea bed of the continental shelf beneath the high seas but contiguous to the coasts of the United States as appertaining to the United States, subject to its jurisdiction and control. In cases where the continental shelf extends to the shores of another State, or is shared with an adjacent State, the boundary shall be determined by the United States and the State concerned in accordance with equitable principles. The character as high seas of the waters above the continental shelf and the right to their free and unimpeded navigation are in no way thus affected.

[seal]   HARRY S. TRUMAN

By the President:

Dean Acheson,

   Acting Secretary of State.

September 28, 1945.

Appendix 3

A Proclamation (No. 2668):
Policy of the United States with
Respect to Coastal Fisheries
in Certain Areas of the High Seas
(By the President of the United States of America)

Whereas for some years the Government of the United States of America has viewed with concern the inadequacy of present arrangements for the protection and perpetuation of the fishery resources contiguous to its coasts, and, in view of the potentially disturbing effect of this situation, has carefully studied the possibility of improving the jurisdictional basis for conservation measures and international cooperation in this field; and

Whereas such fishery resources have a special importance to coastal communities as a source of livelihood and to the nation as a food and industrial resource; and

Whereas the progressive development of new methods and techniques contributes to intensified fishing over wide sea areas and in certain cases seriously threatens fisheries with depletion; and

Whereas there is an urgent need to protect coastal fishery resources from destructive exploitation, having due regard to conditions peculiar to each region and situation and to the special rights and equities of the coastal State and of any other State which may have established a legitimate interest therein;

Now, therefore, I, Harry S. Truman, President of the United States of America, do hereby proclaim the following policy of the United States of America with respect to coastal fisheries in certain areas of the high seas:

In view of the pressing need for conservation and protection of fishery resources, the Government of the United States regards it as proper to establish conservation zones in those areas of the high seas contiguous to the coasts of the United States wherein fishing activities have been or in the future may be developed and maintained on a substantial scale. Where such activities have been or shall hereafter be developed and maintained by its nationals alone, the United States regards it as proper to establish explicitly bounded conservation zones in which fishing activities shall be subject to the regulation and control of the United States. Where such activities have been or shall hereafter be legitimately developed and maintained jointly by nationals of the United States and nationals of other States, explicitly bounded conservation zones may be established under agreements between the United States and such other States; and all fishing activities in such zones shall be subject to regulation and control as provided in such agreements. The right of any State to establish conservation zones off its shores in accordance with the above principles is conceded, provided that corresponding recognition is given to any fishing interests of nationals of the United States which may exist in such areas. The character as high seas of the areas in which such conservation zones are established and the right to their free and unimpeded navigation are in no way thus affected.

IN WITNESS WHEREOF, I have hereunto set my hand and caused the seal of the United States of America to be affixed.

DONE at the City of Washington this twenty-eighth day of September, in the year of our Lord nineteen hundred and forty-five, and of the Independence of the United States of America the one hundred and seventieth.

[seal] HARRY S. TRUMAN

By the President:

Dean Acheson,
  Acting Secretary of State.
September 28, 1945.

Appendix 4

Press Releases Relative to the Natural Resources of the Continental Shelf

THE WHITE HOUSE, *September 28, 1945.*

The President today issued two proclamations asserting the jurisdiction of the United States over the natural resources of the continental shelf under the high seas contiguous to the coasts of the United States and its territories, and providing for the establishment of conservation zones for the protection of fisheries in certain areas of the high seas contiguous to the United States. The action of the President in regard to both the resources of the continental shelf and the conservation of high seas fisheries in which the United States has an interest was taken on the recommendation of the Secretary of State and the Secretary of the Interior.

Two companion Executive orders were also issued by the President. One reserved and set aside the resources of the continental shelf under the high seas and placed them for administrative purposes, pending legislative action, under the jurisdiction and control of the Secretary of the Interior. The other provided for the establishment by Executive orders, on recommendation of the Secretary of State and the Secretary of the Interior, of fishery conservation zones in areas of the high seas contiguous to the coasts of the United States.

Until the present the only high seas fisheries in the regulation of which the United States has participated, under treaties or conventions, are those for whales, Pacific halibut and fur seals.

In areas where fisheries have been or shall hereafter be developed and maintained by nationals of the United States alone, explicitly bounded zones will be set up in which the United States may regulate and control all fishing activities.

In other areas where the nationals of other countries, as well as our own, have developed or shall hereafter legitimately develop fisheries, zones may be established by agreements between the United States and such other States and joint regulations and control will be put into effect.

The United States will recognize the rights of other countries to establish conservation zones off their own coasts where the interests of nationals of the United States are recognized in the same manner that we recognize the interests of the nationals of the other countries.

The assertion of this policy has long been advocated by conservationists, including a substantial section of the fishing industry of the United States, since regulation of a fishery resource within territorial waters cannot control the misuse or prevent the depletion of that resource through uncontrolled fishery activities conducted outside of the commonly accepted limits of territorial jurisdiction.

As a result of the establishment of this new policy, the United States will be able to protect effectively, for instance, its most valuable fishery, that for the Alaska salmon. Through painstaking conservation efforts and scientific management the United States has made excellent progress in maintaining the salmon at high levels. However, since the salmon spends a considerable portion of its life in the open sea, uncontrolled fishery activities on the high seas, either by nationals of the United States or other countries, have constituted an ever present menace to the salmon fishery.

The policy proclaimed by the President in regard to the jurisdiction over the continental shelf does not touch upon the question of Federal versus State control. It is concerned solely with establishing the jurisdiction of the United States from an international standpoint. It will, however, make possible the orderly development of an underwater area 750,000 square miles in extent. Generally, submerged land which is contiguous to the continent and which is covered by no more than 100 fathoms (600 feet) of water is considered as the continental shelf.

Petroleum geologists believe that portions of the continental shelf beyond the 3-mile limit contain valuable oil deposits. The study of subsurface structures associated with oil deposits which have been discovered along the Gulf Coast of Texas, for instance, indicates that corresponding deposits may underlie the offshore or submerged land. The trend of oil-productive salt domes extends directly into the Gulf of Mexico off the Texas coast. Oil is also being taken at present from wells within the 3-mile limit off the coast of California. It is quite possible, geologists say, that the oil deposits extend beyond this traditional limit of national jurisdiction.

Valuable deposits of minerals other than oil may also be expected to be found in these submerged areas. Ore mines now extend under the sea from the coasts of England, Chile, and other countries.

While asserting jurisdiction and control of the United States over the mineral resources of the continental shelf, the proclamation in no wise abridges the right of free and unimpeded navigation of waters of the character of high seas above the shelf, nor does it extend the present limits of the territorial waters of the United States.

The advance of technology prior to the present war had already made possible the exploitation of a limited amount of minerals from submerged lands within the 3-mile limit. The rapid development of technical knowledge and equipment occasioned by the war, now makes possible the determination of the resources of the submerged lands outside of the 3-mile limit. With the need for the discovery of additional resources of petroleum and other minerals it became advisable for the United States to make possible orderly development of these resources. The proclamation of the President is designed to serve this purpose.

Appendix 5

Public Law 31

An act to confirm and establish the titles of the States to lands beneath navigable waters within State boundaries and to the natural resources within such lands and waters, to provide for the use and control of said lands and resources, and to confirm the jurisdiction and control of the United States over the natural resources of the seabed of the Continental Shelf seaward of State boundaries

Be it enacted by the Senate and House of Representatives of the United States of America in Congress assembled, That this Act may be cited as the "Submerged Lands Act."

TITLE I

Definition

Sec. 2. When used in this Act—
(a) The term "lands beneath navigable waters" means—
1. All lands within the boundaries of each of the respective States which are covered by nontidal waters that were navigable under the laws of the United States at the time such State became a member of the Union, or acquired sovereignty over such lands and waters thereafter, up to the ordinary high-water mark as heretofore or hereafter modified by accretion, erosion, and reliction;
2. All lands permanently or periodically covered by tidal waters up to but not above the line of mean high tide and seaward to a line three geographical miles distant from the coast line of each such State and to the boundary line of each such State where in any case such boundary as it existed at the time such State became a member of the Union, or as heretofore approved by Congress, extends seaward (or into the Gulf of Mexico) beyond three geographical miles, and
3. All filled in, made, or reclaimed lands which formerly were lands beneath navigable waters, as hereinabove defined;
(b) The term "boundaries" includes the seaward boundaries of a

State or its boundaries in the Gulf of Mexico or any of the Great Lakes as they existed at the time such State became a member of the Union, or as heretofore approved by the Congress, or as extended or confirmed pursuant to section 4 hereof but in no event shall the term "boundaries" or the term "lands beneath navigable waters" be interpreted as extending from the coast line more than three geographical miles into the Atlantic Ocean or the Pacific Ocean, or more than three marine leagues into the Gulf of Mexico;

(c) The term "coast line" means the line of ordinary low water along that portion of the coast which is in direct contact with the open sea and the line marking the seaward limit of inland waters;

(d) The terms "grantees" and "lessees" include (without limiting the generality thereof) all political subdivisions, municipalities, public and private corporations, and other persons holding grants or leases from a State, or from its predecessor sovereign if legally validated, to lands beneath navigable waters if such grants or leases were issued in accordance with the constitution, statutes, and decisions of the courts of the State in which such lands are situated, or of its predecessor sovereign: *Provided, however,* That nothing herein shall be construed as conferring upon said grantees or lessees any greater rights or interests other than are described herein and in their respective grants from the State, or its predecessor sovereign;

(e) The term "natural resources" includes, without limiting the generality thereof, oil, gas, and all other minerals, and fish, shrimp, oysters, clams, crabs, lobsters, sponges, kelp, and other marine animal and plant life but does not include water power, or the use of water for the production of power;

(f) The term "lands beneath navigable waters" does not include the beds of streams in lands now or heretofore constituting a part of the public lands of the United States if such streams were not meandered in connection with the public survey of such lands under the laws of the United States and if the title to the beds of such streams was lawfully patented or conveyed by the United States or any State to any person;

(g) The term "State" means any State of the Union;

(h) The term "person" includes, in addition to a natural person, an association, a State, a political subdivision of a State, or a private, public, or municipal corporation.

TITLE II

Lands Beneath Navigable Waters Within State Boundaries

Sec. 3. Rights of the States.—

(a) It is hereby determined and declared to be in the public interest that (1) title to and ownership of the lands beneath navigable waters within the boundaries of the respective States, and the natural resources within such lands and waters, and (2) the right and power to manage, administer, lease, develop, and use the said lands and natural resources all in accordance with applicable State law be, and they are hereby, subject to the provisions hereof, recognized, confirmed, established, and vested in and assigned to the respective States or the persons who were on June 5, 1950, entitled thereto under the law of the respective States in which the land is located, and the respective grantees, lessees, or successors in interest thereof;

(b) (1) The United States hereby releases and relinquishes unto said States and persons aforesaid, except as otherwise reserved herein, all right, title, and interest of the United States, if any it has, in and to all said lands, improvements, and natural resources; (2) the United States hereby releases and relinquishes all claims of the United States, if any it has, for money or damages arising out of any operations of said States or persons pursuant to State authority upon or within said lands and navigable waters; and (3) the Secretary of the Interior or the Secretary of the Navy or the Treasurer of the United States shall pay to the respective States or their grantees issuing leases covering such lands or natural resources all moneys paid thereunder to the Secretary of the Interior or to the Secretary of the Navy or to the Treasurer of the United States and subject to the control of any of them or to the control of the United States on the effective date of this Act, except that portion of such moneys which (1) is required to be returned to a lessee; or (2) is deductible as provided by stipulation or agreement between the United States and any of said States;

(c) The rights, powers, and titles hereby recognized, confirmed, established, and vested in and assigned to the respective States and their grantees are subject to each lease executed by a State, or its grantee, which was in force and effect on June 5, 1950, in accordance with its terms and provisions and the laws of the State issuing, or whose grantee issued, such lease, and such rights, powers, and titles are further subject to the rights herein now granted to any person holding any such lease to continue to maintain the lease, and to conduct operations thereunder, in accordance with its provisions, for the full term thereof, and any extensions, renewals, or replacements authorized therein, or heretofore authorized by the laws of the State issuing, or whose grantee issued such lease: *Provided, however,* That, if oil or gas was not being produced from such lease on and before December 11, 1950, or if the primary term of such lease has expired since December 11, 1950, then for a term from the effective date hereof equal to the term remaining unexpired on December 11, 1950, under the provisions of such lease or any extensions, renewals, or replacements authorized therein or heretofore authorized by the laws of the State issuing, or whose grantee issued, such lease: *Provided, however,* That within ninety days from the effective date hereof (i) the lessee shall pay to the State or its grantee issuing such lease all rents, royalties, and other sums payable between June 5,

1950, and the effective date hereof, under such lease and the laws of the State issuing or whose grantee issued such lease, except such rents, royalties, and other sums as have been paid to the State, its grantee, the Secretary of the Interior or the Secretary of the Navy or the Treasurer of the United States and not refunded to the lessee; and (ii) the lessee shall file with the Secretary of the Interior or the Secretary of the Navy and with the State issuing or whose grantee issued such lease, instruments consenting to the payment by the Secretary of the Interior or the Secretary of the Navy or the Treasurer of the United States to the State or its grantee issuing the lease, of all rents, royalties, and other payments under the control of the Secretary of the Interior or the Secretary of the Navy or the Treasurer of the United States or the United States which have been paid, under the lease, except such rentals, royalties, and other payments as have also been paid by the lessee to the State or its grantee;

(d) Nothing in this Act shall affect the use, development, improvement, or control by or under the constitutional authority of the United States of said lands and waters for the purposes of navigation or flood control or the production of power, or be construed as the release or relinquishment of any rights of the United States arising under the constitutional authority of Congress to regulate or improve navigation, or to provide for flood control, or the production of power;

(e) Nothing in this Act shall be construed as affecting or intended to affect or in any way interfere with or modify the laws of the States which lie wholly or in part westward of the ninety-eighth meridian, relating to the ownership and control of ground and surface waters; and the control, appropriation, use, and distribution of such waters shall continue to be in accordance with the laws of such States.

Sec. 4. Seaward Boundaries.—The seaward boundary of each original coastal State is hereby approved and confirmed as a line three geographical miles distant from its coast line or, in the case of the Great Lakes, to the international boundary. Any State admitted subsequent to the formation of the Union which has not already done so may extend its seaward boundaries to a line three geographical miles distant from its coast line, or to the international boundaries of the United States in the Great Lakes or any other body of water traversed by such boundaries. Any claim heretofore or hereafter asserted either by constitutional provision, statute, or otherwise, indicating the intent of a State so to extend its boundaries is hereby approved and confirmed, without prejudice to its claim, if any it has, that its boundaries extend beyond that line. Nothing in this section is to be construed as questioning or in any manner prejudicing the existence of any State's seaward boundary beyond three geographical miles if it was so provided by its constitution or laws prior to or at the time such State became a member of the Union, or if it has been heretofore approved by Congress.

Sec. 5. Exceptions From Operation of Section 3 of This Act.—There is excepted from the operation of section 3 of this Act—

(a) all tracts or parcels of land together with all accretions thereto, resources therein, or improvements thereon, title to which has been lawfully and expressly acquired by the United States from any State or from any person in whom title had vested under the law of the State or of the United States, and all lands which the United States lawfully holds under the law of the State; all lands expressly retained by or ceded to the United States when the State entered the Union (otherwise than by a general retention or cession of lands underlying the marginal sea); all lands acquired by the United States by eminent domain proceedings, purchase, cession, gift, or otherwise in a proprietary capacity; all lands filled in, built up, or otherwise reclaimed by the United States for its own use; and any rights the United States has in lands presently and actually occupied by the United States under claim of right;

(b) such lands beneath navigable waters held, or any interest in which is held by the United States for the benefit of any tribe, band, or group of Indians or for individual Indians; and

(c) all structures and improvements constructed by the United States in the exercise of its navigational servitude.

Sec. 6. Powers Retained by the United States.—(a) The United States retains all its navigational servitude and rights in and powers of regulation and control of said lands and navigable waters for the constitutional purposes of commerce, navigation, national defense, and international affairs, all of which shall be paramount to, but shall not be deemed to include, proprietary rights of ownership, or the rights of management, administration, leasing, use, and development of the lands and natural resources which are specifically recognized, confirmed, established, and vested in and assigned to the respective States and others by section 3 of this Act.

(b) In time of war or when necessary for national defense, and the Congress or the President shall so prescribe, the United States shall have the right of first refusal to purchase at the prevailing market price, all or any portion of the said natural resources, or to acquire and use any portion of said lands by proceeding in accordance with due process of law and paying just compensation therefor.

Sec. 7. Nothing in this Act shall be deemed to amend, modify, or repeal the Acts of July 26, 1866 (14 Stat. 251), July 9, 1870 (16 Stat. 217), March 3, 1877 (19 Stat. 377), June 17, 1902 (32 Stat. 388), and

December 22, 1944 (58 Stat. 887), and Acts amendatory thereof or supplementary thereto.

Sec. 8. Nothing contained in this Act shall affect such rights, if any, as may have been acquired under any law of the United States by any person in lands subject to this Act and such rights, if any, shall be governed by the law in effect at the time they may have been acquired: *Provided, however,* That nothing contained in this Act is intended or shall be construed as a finding, interpretation, or construction by the Congress that the law under which such rights may be claimed in fact or in law applies to the lands subject to this Act, or authorizes or compels the granting of such rights in such lands, and that the determination of the applicability or effect of such law shall be unaffected by anything contained in this Act.

Sec. 9. Nothing in this Act shall be deemed to affect in any wise the rights of the United States to the natural resources of that portion of the subsoil and seabed of the Continental Shelf lying seaward and outside of the area of lands beneath navigable waters, as defined in section 2 hereof, all of which natural resources appertain to the United States, and the jurisdiction and control of which by the United States is hereby confirmed.

Sec. 10. Executive Order Numbered 10426, dated January 16, 1953, entitled "Setting Aside Submerged Lands of the Continental Shelf as a Naval Petroleum Reserve," is hereby revoked insofar as it applies to any lands beneath navigable waters as defined in section 2 hereof.

Sec. 11. Separability.—If any provision of this Act, or any section, subsection, sentence, clause, phrase or individual word, or the application thereof to any person or circumstance is held invalid, the validity of the remainder of the Act and of the application of any such provision, section, subsection, sentence, clause, phrase or individual word to other persons and circumstances shall not be affected thereby; without limiting the generality of the foregoing, if subsection 3 (a) 1, 3 (a) 2, 3 (b) 1, 3 (b) 2, 3 (b) 3, or 3, or 3 (c) or any provision of any of those subsections is held invalid such subsection or provision shall be held separable and the remaining subsections and provisions shall not be affected thereby.

Approved May 22, 1953.

Appendix 6

Executive Order 9633:
Reserving and Placing Certain Resources of the Continental Shelf Under the Control and Jurisdiction of the Secretary of the Interior

By virtue of and pursuant to the authority vested in me as President of the United States, it is ordered that the natural resources of the subsoil and seabed of the Continental Shelf beneath the high seas but contiguous to the coasts of the United States declare this day by proclamation to appertain to the United States and to be subject to its jurisdiction and control, be and they are hereby reserved, set aside, and placed under the jurisdiction and control of the Secretary of the Interior for administrative purposes pending the enactment of legislation in regard thereto. Neither this order nor the aforesaid proclamation shall be deemed to affect the determination by legislation or judicial decree of any issues between the United States and the several States relating to the ownership or control of the subsoil and seabed of the Continental Shelf within or outside of the 3-mile limit.

HARRY S. TRUMAN

The White House, *September 28, 1945.*

Appendix 7

Executive Order 10426:
Setting Aside Submerged Lands of the
Continental Shelf as a Naval Petroleum Reserve

By virtue of the authority vested in me as President of the United States, it is ordered as follows:

Sec. 1. (a) Subject to valid existing rights, if any, and to the provisions of this order, the lands of the continental shelf of the United States and Alaska lying seaward of the line of mean low tide and outside the inland waters and extending to the furthermost limits of the paramount rights, full dominion, and power of the United States over lands of the continental shelf are hereby set aside as a naval petroleum reserve and shall be administered by the Secretary of the Navy.

(b) The reservation established by this section shall be for oil and gas only, and shall not interfere with the use of the lands or waters within the reserved area for any lawful purpose not inconsistent with the reservation.

Sec. 2. The provisions of this order shall not affect the operating stipulation which was entered into on July 26, 1947, by the Attorney General of the United States and the Attorney General of California in the case of *United States of America* v. *State of California* (in the Supreme Court of the United States, October Term, 1947, No. 12, Original), as thereafter extended and modified.

Sec. 3. (a) The functions of the Secretary of the Interior under Parts II and III of the notice issued by the Secretary of the Interior on December 11, 1950, and entitled "Oil and Gas Operations in the Submerged Coastal Lands of the Gulf of Mexico" (15 F. R. 8835), as supplemented and amended, are transferred to the Secretary of the Navy; and the term "Secretary of the Navy" shall be substituted for the term "Secretary of the Interior" wherever the latter term occurs in the said Parts II and III.

(b) Paragraph (c) of Part III of the aforesaid notice dated December 11, 1950, as amended, is amended to read as follows:

"(c) The remittance shall be deposited in a suspense account within the Treasury of the United States, subject to the control of the Secretary of the Navy, the proceeds to be expended in such manner as may hereafter be directed by an act of Congress or, in the absence of such direction, refunded (which may include a refund of the money for reasons other than those hereinafter set forth) or deposited into the general fund of the Treasury, as the Secretary of the Navy may deem to be proper."

(c) The provisions of Parts II and III of the aforesaid notice dated December 11, 1950, as supplemented and amended, including the amendments made by this order, shall continue in effect until changed by the Secretary of the Navy.

Sec. 4. Executive Order No. 9633 of September 28, 1945, entitled "Reserving and Placing Certain Resources of the Continental Shelf Under the Control and Jurisdiction of the Secretary of the Interior" (10 F. R. 12305), is hereby revoked.

HARRY S. TRUMAN

The White House, *January 16, 1953.*

(F. R. Doc. 53–734; Filed, Jan. 16, 1953; 4:56 p.m.)

Appendix 8

Public Law 212

An Act to provide for the jurisdiction of the United States over the submerged lands of the outer Continental Shelf, and to authorize the Secretary of the Interior to lease such lands for certain purposes

Be it enacted by the Senate and House of Representatives of the United States of America in Congress assembled, That this Act may be cited as the "Outer Continental Shelf Lands Act."

Sec. 2. Definitions.—When used in this Act—

(a) The term "outer Continental Shelf" means all submerged lands lying seaward and outside of the area of lands beneath navigable waters as defined in section 2 of the Submerged Lands Act (Public Law 31, Eighty-third Congress, first session), and of which the subsoil and seabed appertain to the United States and are subject to its jurisdiction and control;

(b) The term "Secretary" means the Secretary of the Interior;

(c) The term "mineral lease" means any form of authorization for the exploration for, or development or removal of deposits of, oil, gas, or other minerals; and

(d) The term "person" includes, in addition to a natural person, an association, a State, a political subdivision of a State, or a private, public, or municipal corporation.

Sec. 3. Jurisdiction Over Outer Continental Shelf.—(a) It is hereby declared to be the policy of the United States that the subsoil and seabed of the outer Continental Shelf appertain to the United States and are subject to its jurisdiction, control, and power of disposition as provided in this Act.

(b) This Act shall be construed in such manner that the character as high seas of the waters above the outer Continental Shelf and the right to navigation and fishing therein shall not be affected.

Sec. 4. Laws Applicable to Outer Continental Shelf.—(a) (1) The Constitution and laws and civil and political jurisdiction of the United States are hereby extended to the subsoil and seabed of the outer Continental Shelf and to all artificial islands and fixed structures which may be erected thereon for the purpose of exploring for, developing, removing, and transporting resources therefrom, to the same extent as if the outer Continental Shelf were an area of exclusive Federal jurisdiction located within a State: *Provided, however,* That mineral leases on the outer Continental Shelf shall be maintained or issued only under the provisions of this Act.

(2) To the extent that they are applicable and not inconsistent with this Act or with other Federal laws and regulations of the Secretary now in effect or hereafter adopted, the civil and criminal laws of each adjacent State as of the effective date of this Act are hereby declared to be the law of the United States for that portion of the subsoil and seabed of the outer Continental Shelf, and artificial islands and fixed structures erected thereon, which would be within the area of the State if its boundaries were extended seaward to the outer margin of the outer Continental Shelf, and the President shall determine and publish in the Federal Register such projected lines extending seaward and defining each such area. All of such applicable laws shall be administered and enforced by the appropriate officers and courts of the United States. State taxation laws shall not apply to the outer Continental Shelf.

(3) The provisions of this section for adoption of State law as the law of the United States shall never be interpreted as a basis for claiming any interest in or jurisdiction on behalf of any State for any purpose over the seabed and subsoil of the outer Continental Shelf, or the property and natural resources thereof or the revenues therefrom.

(b) The United States district courts shall have original jurisdiction of cases and controversies arising out of or in connection with any operations conducted on the outer Continental Shelf for the purpose of exploring for, developing, removing or transporting by pipeline the natural resources, or involving rights to the natural resources of the subsoil and seabed of the outer Continental Shelf, and proceedings with respect to any such case or controversy may be instituted in the judicial district in which any defendant resides or may be found, or in the judicial district of the adjacent State nearest the place where the cause of action arose.

(c) With respect to disability or death of an employee resulting from any injury occurring as the result of operations described in subsection (b), compensation shall be payable under the provisions of the Longshoremen's and Harbor Workers' Compensation Act. For the purposes of the extension of the provisions of the Longshoremen's and Harbor Workers' Compensation Act under this section—

1. The term "employee" does not include a master or member of a crew of any vessel, or an officer or employee of the United States or any agency thereof or of any State or foreign government, or of any political subdivision thereof;

2. The term "employer" means an employer any of whose employees are employed in such operations; and

3. the term "United States" when used in a geographical sense includes the outer Continental Shelf and artificial islands and fixed structures thereon.

(d) For the purposes of the National Labor Relations Act, as amended, any unfair labor practice, as defined in such Act, occurring upon any artificial island or fixed structure referred to in subsection (a) shall be deemed to have occurred within the judicial district of the adjacent State nearest the place of location of such island or structure.

(e) (1) The head of the Department in which the Coast Guard is operating shall have authority to promulgate and enforce such reasonable regulations with respect to lights and other warning devices, safety equipment, and other matters relating to the promotion of safety of life and property on the islands and structures referred to in subsection (a) or on the waters adjacent thereto, as he may deem necessary.

(2) The head of the Department in which the Coast Guard is operating may mark for the protection of navigation any such island or structure whenever the owner has failed suitably to mark the same in accordance with regulations issued hereunder, and the owner shall pay the cost thereof. Any person, firm, company, or corporation who shall fail or refuse to obey any of the lawful rules and regulations issued hereunder shall be guilty of a misdemeanor and shall be fined not more than $100 for each offense. Each day during which such violation shall continue shall be considered a new offense.

(f) The authority of the Secretary of the Army to prevent obstruction to navigation in the navigable waters of the United States is hereby extended to artificial islands and fixed structures located on the outer Continental Shelf.

(g) The specific application by this section of certain provisions of law to the subsoil and seabed of the outer Continental Shelf and the artificial islands and fixed structures referred to in subsection (a) or to acts or offenses occurring or committed thereon shall not give rise to any inference that the application to such islands and structures, acts, or offenses of any other provision of law is not intended.

Sec. 5. Administration of Leasing of the Outer Continental Shelf.— (a) (1) The Secretary shall administer the provisions of this Act relating to the leasing of the outer Continental Shelf, and shall prescribe such rules and regulations as may be necessary to carry out such provisions. The Secretary may at any time prescribe and amend such rules and regulations as he determines to be necessary and proper in order to provide for the prevention of waste and conservation of the natural resources of the outer Continental

Shelf, and the protection of correlative rights therein, and, notwithstanding any other provisions herein, such rules and regulations shall apply to all operations conducted under a lease issued or maintained under the provisions of this Act. In the enforcement of conservation laws, rules, and regulations the Secretary is authorized to cooperate with the conservation agencies of the adjacent States. Without limiting the generality of the foregoing provisions of this section, the rules and regulations prescribed by the Secretary thereunder may provide for the assignment or relinquishment of leases, for the sale of royalty oil and gas accruing or reserved to the United States at not less than market value, and, in the interest of conservation, for unitization, pooling, drilling agreements, suspension of operations or production, reduction of rentals or royalties, compensatory royalty agreements, subsurface storage of oil or gas in any of said submerged lands, and drilling or other easements necessary for operations or production.

(2) Any person who knowingly and willfully violates any rule or regulation prescribed by the Secretary for the prevention of waste, the conservation of the natural resources, or the protection of correlative rights shall be deemed guilty of a misdemeanor and punishable by a fine of not more than $2,000 or by imprisonment for not more than six months, or by both such fine and imprisonment, and each day of violation shall be deemed to be a separate offense. The issuance and continuance in effect of any lease, or of any extension, renewal, or replacement of any lease under the provisions of this Act shall be conditioned upon compliance with the regulations issued under this Act and in force and effect on the date of the issuance of the lease if the lease is issued under the provisions of section 8 hereof, or with the regulations issued under the provisions of section 6(b), clause (2), hereof if the lease is maintained under the provisions of section 6 hereof.

(b) (1) Whenever the owner of a nonproducing lease fails to comply with any of the provisions of this Act, or of the lease, or of the regulations issued under this Act and in force and effect on the date of the issuance of the lease if the lease is issued under the provisions of section 8 hereof, or of the regulations issued under the provisions of section 6(b), clause (2), hereof, if the lease is maintained under the provisions of section 6 hereof, such lease may be canceled by the Secretary, subject to the right of judicial review as provided in section 8(j), if such default continues for the period of thirty days after mailing of notice by registered letter to the lease owner at his record post office address.

(2) Whenever the owner of any producing lease fails to comply with any of the provisions of this Act, or of the lease, or of the regulations issued under this Act and in force and effect on the date of the issuance of the lease if the lease is issued under the provisions of section 8 hereof, or of the regulations issued under the provisions of section 6(b), clause (2), hereof,

if the lease is maintained under the provisions of section 6 hereof, such lease may be forfeited and canceled by an appropriate proceeding in any United States district court having jurisdiction under the provisions of section 4(b) of this Act.

(c) Rights-of-way through the submerged lands of the outer Continental Shelf, whether or not such lands are included in a lease maintained or issued pursuant to this Act, may be granted by the Secretary for pipeline purposes for the transportation of oil, natural gas, sulphur, or other mineral under such regulations and upon such conditions as to the application therefor and the survey, location and width thereof as may be prescribed by the Secretary, and upon the express condition that such oil or gas pipelines shall transport or purchase without discrimination, oil or natural gas produced from said submerged lands in the vicinity of the pipeline in such proportionate amounts as the Federal Power Commission, in the case of gas, and the Interstate Commerce Commission, in the case of oil, may, after a full hearing with due notice thereof to the interested parties, determine to be reasonable, taking into account, among other things, conservation and the prevention of waste. Failure to comply with the provisions of this section or the regulations and conditions prescribed thereunder shall be ground for forfeiture of the grant in an appropriate judicial proceeding instituted by the United States in any United States district court having jurisdiction under the provisions of section 4 (b) of this Act.

Sec. 6. Maintenance of Leases on Outer Continental Shelf.—(a) The provisions of this section shall apply to any mineral lease covering submerged lands of the outer Continental Shelf issued by any State (including any extension, renewal, or replacement thereof heretofore granted pursuant to such lease or under the laws of such State) if—

1. Such lease, or a true copy thereof, is filed with the Secretary by the lessee or his duly authorized agent within ninety days from the effective date of this Act, or within such further period or periods as provided in section 7 hereof or as may be fixed from time to time by the Secretary;

2. Such lease was issued prior to December 21, 1948, and would have been on June 5, 1950, in force and effect in accordance with its terms and provisions and the law of the State issuing it had the State had authority to issue such lease;

3. There is filed with the Secretary, within the period or periods specified in paragraph (1) of this subsection, (A) a certificate issued by the State official or agency having jurisdiction over such lease stating that it would have been in force and effect as required by the provisions of paragraph (2) of this subsection, or (B) in the absence of such certificate, evidence in the form of affidavits, receipts, canceled checks, or other documents that may

be required by the Secretary, sufficient to prove that such lease would have been so in force and effect;

4. Except as otherwise provided in section 7 hereof, all rents, royalties, and other sums payable under such lease between June 5, 1950, and the effective date of this Act, which have not been paid in accordance with the provisions thereof, or to the Secretary or to the Secretary of the Navy, are paid to the Secretary within the period or periods specified in paragraph (1) of this subsection and all rents, royalties, and other sums payable under such lease after the effective date of this Act, are paid to the Secretary, who shall deposit such payments in the Treasury in accordance with section 9 of this Act;

5. The holder of such lease certifies that such lease shall continue to be subject to the overriding royalty obligations existing on the effective date of this Act;

6. Such lease was not obtained by fraud or misrepresentation;

7. Such lease, if issued on or after June 23, 1947, was issued upon the basis of competitive bidding;

8. Such lease provides for a royalty to the lessor on oil and gas of not less than 12½ per centum and on sulphur of not less than 5 per centum in amount or value of the production saved, removed, or sold from the lease, or, in any case in which the lease provides for a lesser royalty, the holder thereof consents in writing, filed with the Secretary, to the increase of the royalty to the minimum herein specified;

9. The holder thereof pays to the Secretary within the period or periods specified in paragraph (1) of this subsection an amount equivalent to any severance, gross production, or occupation taxes imposed by the State issuing the lease on the production from the lease, less the State's royalty interest in such production, between June 5, 1950, and the effective date of this Act and not heretofore paid to the State, and thereafter pays to the Secretary as an additional royalty on the production from the lease, less the United States' royalty interest in such production, a sum of money equal to the amount of the severance, gross production, or occupation taxes which would have been payable on such production to the State issuing the lease under its laws as they existed on the effective date of this Act;

10. Such lease will terminate within a period of not more than five years from the effective date of this Act in the absence of production or operations for drilling, or, in any case in which the lease provides for a longer period, the holder thereof consents in writing, filed with the Secretary, to the reduction of such period so that it will not exceed the maximum period herein specified; and

11. The holder of such lease furnishes such surety bond, if any, as the Secretary may require and complies with such other reasonable requirements as the Secretary may deem necessary to protect the interests of the United States.

(b) Any person holding a mineral lease, which as determined by the Secretary meets the requirements of subsection (a) of this section, may continue to maintain such lease, and may conduct operations thereunder, in accordance with (1) its provisions as to the area, the minerals covered, rentals and, subject to the provisions of paragraphs (8), (9) and (10) of subsection (a) of this section, as to royalties and as to the term thereof and of any extensions, renewals, or replacements authorized therein or heretofore authorized by the laws of the State issuing such lease, or, if oil or gas was not being produced in paying quantities from such lease on or before December 11, 1950, or if production in paying quantities has ceased since June 5, 1950, or if the primary term of such lease has expired since December 11, 1950, then for a term from the effective date hereof equal to the term remaining unexpired on December 11, 1950, under the provisions of such lease or any extensions, renewals, or replacements authorized therein, or heretofore authorized by the laws of such State, and (2) such regulations as the Secretary may under section 5 of this Act prescribe within ninety days after making his determination that such lease meets the requirements of subsection (a) of this section: *Provided, however,* That any rights to sulphur under any lease maintained under the provisions of this subsection shall not extend beyond the primary term of such lease or any extension thereof under the provisions of such subsection (b) unless sulphur is being produced in paying quantities or drilling, well reworking, plant construction, or other operations for the production of sulphur, as approved by the Secretary, are being conducted on the area covered by such lease on the date of expiration of such primary term or extension: *Provided further,* That if sulphur is being produced in paying quantities on such date, then such rights shall continue to be maintained in accordance with such lease and the provisions of this Act: *Provided further,* That, if the primary term of a lease being maintained under subsection (b) hereof has expired prior to the effective date of this Act and oil or gas is being produced in paying quantities on such date, then such rights to sulphur as the lessee may have under such lease shall continue for twenty-four months from the effective date of this Act and as long thereafter as sulphur is produced in paying quantities, or drilling, well working, plant construction, or other operations for the production of sulphur, as approved by the Secretary, are being conducted on the area covered by the lease.

(c) The permission granted in subsection (b) of this section shall not be construed to be a waiver of such claims, if any, as the United States may have against the lessor or the lessee or any other person respecting sums payable or paid for or under the lease, or respecting activities conducted under the lease, prior to the effective date of this Act.

(d) Any person complaining of a negative determination by the Secretary of the Interior under this section may have such determination reviewed by the United States Distict Court for the District of Columbia by filing a petition for review within sixty days after receiving notice of such action by the Secretary.

(e) In the event any lease maintained under this section covers lands beneath navigable waters, as that term is used in the Submerged Lands Act, as well as lands of the outer Continental Shelf, the provisions of this section shall apply to such lease only insofar as it covers lands of the outer Continental Shelf.

Sec. 7. Controversy Over Jurisdiction.—In the event of a controversy between the United States and a State as to whether or not lands are subject to the provisions of this Act, the Secretary is authorized, notwithstanding the provisions of subsections (a) and (b) of section 6 of this Act, and with the concurrence of the Attorney General of the United States, to negotiate and enter into agreements with the State, its political subdivision or grantee or a lessee thereof, respecting operations under existing mineral leases and payment and impounding of rents, royalties, and other sums payable thereunder, or with the State, its political subdivision or grantee, respecting the issuance or nonissuance of new mineral leases pending the settlement or adjudication of the controversy. The authorization contained in the preceding sentence of this section shall not be construed to be a limitation upon the authority conferred on the Secretary in other sections of this Act. Payments made pursuant to such agreement, or pursuant to any stipulation between the United States and a State, shall be considered as compliance with section 6 (a) (4) hereof. Upon the termination of such agreement or stipulation by reason of the final settlement or adjudication of such controversy, if the lands subject to any mineral lease are determined to be in whole or in part lands subject to the provisions of this Act, the lessee, if he has not already done so, shall comply with the requirements of section 6 (a), and thereupon the provisions of section 6 (b) shall govern such lease. The notice concerning "Oil and Gas Operations in the Submerged Coastal Lands of the Gulf of Mexico" issued by the Secretary on December 11, 1950 (15 F. R. 8835), as amended by the notice dated January 26, 1951 (16 F. R. 953), and as supplemented by the notices dated February 2, 1951 (16 F. R. 1203), March 5, 1951 (16 F. R. 2195), April 23, 1951 (16 F. R. 3623), June 25, 1951 (16 F. R. 6404), August 22, 1951 (16 F. R. 8720), October 24, 1951 (16 F. R. 10998), December 21, 1951 (17 F. R. 43), March 25, 1952 (17 F. R. 2821), June 26, 1952 (17 F. R. 5833), and December 24, 1952 (18 F. R. 48), respectively, is hereby approved and confirmed.

Sec. 8. Leasing of Outer Continental Shelf.—(a) In order to meet the urgent need for further exploration and development of the oil and gas deposits of the submerged lands of the outer Continental Shelf, the Secretary is authorized to grant to the highest responsible qualified bidder by competitive bidding under regulations promulgated in advance, oil and gas leases on submerged lands of the outer Continental Shelf which are not covered by leases meeting the requirements of subsection (a) of section 6 of this Act. The bidding shall be (1) by sealed bids, and (2) at the discretion of the Secretary, on the basis of a cash bonus with a royalty fixed by the Secretary at not less than 12½ per centum in amount or value of the production saved, removed or sold, or on the basis of royalty, but at not less than the per centum above mentioned, with a cash bonus fixed by the Secretary.

(b) An oil and gas lease issued by the Secretary pursuant to this section shall (1) cover a compact area not exceeding five thousand seven hundred and sixty acres, as the Secretary may determine, (2) be for a period of five years and as long thereafter as oil or gas may be produced from the area in paying quantities, or drilling or well reworking operations as approved by the Secretary are conducted thereon, (3) require the payment of a royalty of not less than 12½ per centum, in the amount or value of the production saved, removed, or sold from the lease, and (4) contain such rental provisions and such other terms and provisions as the Secretary may prescribe at the time of offering the area for lease.

(c) In order to meet the urgent need for further exploration and development of the sulphur deposits in the submerged lands of the outer Continental Shelf, the Secretary is authorized to grant to the qualified persons offering the highest cash bonuses on a basis of competitive bidding sulphur leases on submerged lands of the outer Continental Shelf, which are not covered by leases which include sulphur and meet the requirements of subsection (a) of section 6 of this Act, and which sulphur leases shall be offered for bid by sealed bids and granted on separate leases from oil and gas leases, and for a separate consideration, and without priority or preference accorded to oil and gas lessees on the same area.

(d) A sulphur lease issued by the Secretary pursuant to this section shall (1) cover an area of such size and dimensions as the Secretary may determine, (2) be for a period of not more than ten years and so long thereafter as sulphur may be produced from the area in paying quantities or drilling, well reworking, plant construction, or other operations for the production of sulphur, as approved by the Secretary, are conducted thereon, (3) require the payment to the United States of such royalty as may be specified in the lease but not less than 5 per centum of the gross production or value of the sulphur at the wellhead, and (4) contain such rental provisions and such other terms and provisions as the Secretary may by regulation prescribe at the time of offering the area for lease.

(e) The Secretary is authorized to grant to the qualified persons offering the highest cash bonuses on a basis of competitive bidding leases of any mineral other than oil, gas and sulphur in any area of the outer Continental Shelf not then under lease for such mineral upon such royalty, rental, and other terms and conditions as the Secretary may prescribe at the time of offering the area for lease.

(f) Notice of sale of leases, and the terms of bidding, authorized by this section shall be published at least thirty days before the date of sale in accordance with rules and regulations promulgated by the Secretary.

(g) All moneys paid to the Secretary for or under leases granted pursuant to this section shall be deposited in the Treasury in accordance with section 9 of this Act.

(h) The issuance of any lease by the Secretary pursuant to this Act, or the making of any interim arrangements by the Secretary pursuant to section 7 of this Act shall not prejudice the ultimate settlement or adjudication of the question as to whether or not the area involved is in the outer Continental Shelf.

(i) The Secretary may cancel any lease obtained by fraud or misrepresentation.

(j) Any person complaining of a cancellation of a lease by the Secretary may have the Secretary's action reviewed in the United States District Court for the District of Columbia by filing a petition for review within sixty days after the Secretary takes such action.

Sec. 9. Disposition of Revenues.—All rentals, royalties, and other sums paid to the Secretary or the Secretary of the Navy under any lease on the outer Continental Shelf for the period from June 5, 1950, to date, and thereafter shall be deposited in the Treasury of the United States and credited to miscellaneous receipts.

Sec. 10. Refunds.—(a) Subject to the provisions of subsection (b) hereof, when it appears to the satisfaction of the Secretary that any person has made a payment to the United States in connection with any lease under this Act in excess of the amount he was lawfully required to pay, such excess shall be repaid without interest to such person or his legal representative, if a request for repayment of such excess is filed with the Secretary within two years after the making of the payment, or within ninety days after the effective date of this Act. The Secretary shall certify the amounts of all such repayments to the Secretary of the Treasury, who is authorized and directed to make such repayments out of any money in the special account established under section 9 of this Act and to issue his warrant in settlement thereof.

(b) No refund of or credit for such excess payment shall be made until after the expiration of thirty days from the date upon which a report giving the name of the person to whom the refund or credit is to be made, the amount of such refund or credit, and a summary of the facts upon which the determination of the Secretary was made is submitted to the President of the Senate and the Speaker of the House of Representatives for transmittal to the appropriate legislative committee of each body, respectively: *Provided,* That if the Congress shall not be in session on the date of such submission or shall adjourn prior to the expiration of thirty days from the date of such submission, then such payment or credit shall not be made until thirty days after the opening day of the next succeeding session of Congress.

Sec. 11. Geological and Geophysical Explorations.—Any agency of the United States and any person authorized by the Secretary may conduct geological and geophysical explorations in the outer Continental Shelf, which do not interfere with or endanger actual operations under any lease maintained or granted pursuant to this Act, and which are not unduly harmful to aquatic life in such area.

Sec. 12. Reservations.—(a) The President of the United States may, from time to time, withdraw from disposition any of the unleased lands of the outer Continental Shelf.

(b) In time of war, or when the President shall so prescribe, the United States shall have the right of first refusal to purchase at the market price all or any portion of any mineral produced from the outer Continental Shelf.

(c) All leases issued under this Act, and leases the maintenance and operation of which are authorized under this Act, shall contain or be construed to contain a provision whereby authority is vested in the Secretary, upon a recommendation of the Secretary of Defense, during a state of war or national emergency declared by the Congress or the President of the United States after the effective date of this Act, to suspend operations under any lease; and all such leases shall contain or be construed to contain provisions for the payment of just compensation to the lessee whose operations are thus suspended.

(d) The United States reserves and retains the right to designate by and through the Secretary of Defense, with the approval of the President, as areas restricted from exploration and operation that part of the outer Continental Shelf needed for national defense; and so long as such designation remains in effect no exploration or operations may be conducted on any part of the surface of such area except with the concurrence of the Secretary of Defense; and if operations or production under any lease theretofore issued on lands within any such restricted area shall be suspended, any payment of rentals, minimum royalty, and royalty prescribed by such lease likewise shall be suspended during such period of suspension of operation and production, and the term of such lease shall be extended by adding thereto any such suspension period, and the United States shall be liable to the lessee for such compensation as is required to be paid under the Constitution of the United States.

(e) All uranium, thorium, and all other materials determined pursuant to paragraph (1) of subsection (b) of section 5 of the Atomic Energy Act of 1946, as amended, to be peculiarly essential to the production of fissionable material, contained, in whatever concentration, in deposits in the subsoil or seabed of the outer Continental Shelf are hereby reserved for the use of the United States.

(f) The United States reserves and retains the ownership of and the right to extract all helium, under such rules and regulations as shall be prescribed by the Secretary, contained in gas produced from any portion of the outer Continental Shelf which may be subject to any lease maintained or granted pursuant to this Act, but the helium shall be extracted from such gas as to cause no substantial delay in the delivery of gas produced to the purchaser of such gas.

Sec. 13. Naval Petroleum Reserve Executive Order Repealed.—Executive Order Numbered 10426, dated January 16, 1953, entitled "Setting Aside Submerged Lands of the Continental Shelf as a Naval Petroleum Reserve," is hereby revoked.

Sec. 14. Prior Claims Not Affected.—Nothing herein contained shall affect such rights, if any, as may have been acquired under any law of the United States by any person in lands subject to this Act and such rights, if any, shall be governed by the law in effect at the time they may have been acquired: *Provided, however,* That nothing herein contained is intended or shall be construed as a finding, interpretation, or construction by the Congress that the law under which such rights may be claimed in fact applies to the lands subject to this Act or authorizes or compels the granting of such rights in such lands, and that the determination of the applicability or effect of such law shall be unaffected by anything herein contained.

Sec. 15. Report by Secretary.—As soon as practicable after the end of each fiscal year, the Secretary shall submit to the President of the Senate and the Speaker of the House of Representatives a report detailing the amounts of all moneys received and expended in connection with the administration of this Act during the preceding fiscal year.

Sec. 16. Appropriations.—There is hereby authorized to be appropriated such sums as may be necessary to carry out the provisions of this Act.

Sec. 17. Separability.—If any provision of this Act, or any section, subsection, sentence, clause, phrase or individual word, or the application thereof to any person or circumstance is held invalid, the validity of the remainder of the Act and of the application of any such provision, section, subsection, sentence, clause, phrase or individual word to other persons and circumstances shall not be affected thereby.

Approved August 7, 1953.

Appendix 9

Conventions on the Law of the Sea Adopted by United Nations Conference at Geneva, 1958

(All conventions contain similar procedural articles for signing, ratification, and revision. They come into force on the thirtieth day following the deposit of the twenty-second instrument of ratification or accession with the United Nations. In the following reprint of the conventions these procedural articles are omitted.)

A. CONVENTION ON THE TERRITORIAL SEA AND THE CONTIGUOUS ZONE*

The States Parties to this Convention, Have agreed as follows:

Part I: Territorial Sea

Section I. General

Article 1

1. The sovereignty of a State extends beyond its land territory and its internal waters, to a belt of sea adjacent to its coast, described as the territorial sea.

2. This sovereignty is exercised subject to the provisions of these articles and to other rules of international law.

Article 2

The sovereignty of a coastal State extends to the air space over the territorial sea as well as to its bed and subsoil.

* Adopted Apr. 27, 1958 (U.N. Doc. A/Conf. 13/L.52).

Section II. Limits of the Territorial Sea

Article 3

Except where otherwise provided in these articles, the normal baseline for measuring the breadth of the territorial sea is the low-water line along the coast as marked on large-scale charts officially recognized by the coastal State.

Article 4

1. In localities where the coastline is deeply indented and cut into, or if there is a fringe of islands along the coast in its immediate vicinity, the method of straight baselines joining appropriate points may be employed in drawing the baseline from which the breadth of the territorial sea is measured.

2. The drawing of such baselines must not depart to any appreciable extent from the general direction of the coast, and the sea areas lying within the lines must be sufficiently closely linked to the land domain to be subject to the régime of internal waters.

3. Baselines shall not be drawn to and from low-tide elevations, unless lighthouses or similar installations which are permanently above sea level have been built on them.

4. Where the method of straight baselines is applicable under the provisions of paragraph 1, account may be taken, in determining particular baselines, of economic interests peculiar to the region concerned, the reality and the importance of which are clearly evidenced by a long usage.

5. The system of straight baselines may not be applied by a State in such a manner as to cut off from the high seas the territorial sea of another State.

6. The coastal State must clearly indicate straight baselines on charts, to which due publicity must be given.

Article 5

1. Waters on the landward side of the baseline of the territorial sea form part of the internal waters of the State.

2. Where the establishment of a straight baseline in accordance with article 4 has the effect of enclosing as internal waters areas which previously had been considered as part of the territorial sea or of the high seas, a right of innocent passage, as provided in articles 14 to 23, shall exist in those waters.

Article 6

The outer limit of the territorial sea is the line every point of which is at a distance from the nearest point of the baseline equal to the breadth of the territorial sea.

Article 7

1. This article relates only to bays the coasts of which belong to a single State.

2. For the purposes of these articles, a bay is a well-marked indentation whose penetration is in such proportion to the width of its mouth as to contain landlocked waters and constitute more than a mere curvature of the coast. An indentation shall not, however, be regarded as a bay unless its area is as large as, or larger than, that of the semicircle whose diameter is a line drawn across the mouth of that indentation.

3. For the purpose of measurement, the area of an indentation is that lying between the low-water mark around the shore of the indentation and a line joining the low-water marks of its natural entrance points. Where, because of the presence of islands, an indentation has more than one mouth, the semicircle shall be drawn on a line as long as the sum total of the lengths of the lines across the different mouths. Islands within an indentation shall be included as if they were part of the water area of the indentation.

4. If the distance between the low-water marks of the natural entrance points of a bay does not exceed twenty-four miles, a closing line may be drawn between these two low-water marks, and the waters enclosed thereby shall be considered as internal waters.

5. Where the distance between the low-water marks of the natural entrance points of a bay exceeds twenty-four miles, a straight baseline of twenty-four miles shall be drawn within the bay in such a manner as to enclose the maximum area of water that is possible with a line of that length.

6. The foregoing provisions shall not apply to so-called "historic" bays, or in any case where the straight baseline system provided for in article 4 is applied.

Article 8

For the purpose of delimiting the territorial sea, the outermost permanent harbour works which form an integral part of the harbour system shall be regarded as forming part of the coast.

Article 9

Roadsteads which are normally used for the loading, unloading and anchoring of ships, and which would otherwise be situated wholly or partly outside the outer limit of the territorial sea, are included in the territorial sea. The coastal State must clearly demarcate such roadsteads and indicate them on charts together with their boundaries, to which due publicity must be given.

Article 10

1. An island is a naturally formed area of land, surrounded by water, which is above water at high tide.

2. The territorial sea of an island is measured in accordance with the provisions of these articles.

Article 11

1. A low-tide elevation is a naturally formed area of land which is surrounded by and above water at low-tide but submerged at high tide. Where a low-tide elevation is situated wholly or partly at a distance not exceeding the breadth of the territorial sea from the mainland or an island, the low-water line on that elevation may be used as the baseline for measuring the breadth of the territorial sea.

2. Where a low-tide elevation is wholly situated at a distance exceeding the breadth of the territorial sea from the mainland or an island, it has no territorial sea of its own.

Article 12

1. Where the coasts of two States are opposite or adjacent to each other, neither of the two States is entitled, failing agreement between them to the contrary, to extend its territorial sea beyond the median line every point of which is equidistant from the nearest points on the baselines from which the breadth of the territorial seas of each of the two States is measured. The provisions of this paragraph shall not apply, however, where it is necessary by reason of historic title or other special circumstances to delimit the territorial seas of the two States in a way which is at variance with this provision.

2. The line of delimitation between the territorial seas of two States lying opposite to each other or adjacent to each other shall be marked on large-scale charts officially recognized by the coastal States.

Article 13

If a river flows directly into the sea, the baseline shall be a straight line across the mouth of the river between points on the low-tide line of its banks.

Section III. Right of Innocent Passage

SUB-SECTION A. RULES APPLICABLE TO ALL SHIPS

Article 14

1. Subject to the provisions of these articles, ships of all States, whether coastal or not, shall enjoy the right of innocent passage through the territorial sea.

2. Passage means navigation through the territorial sea for the purpose either of traversing that sea without entering internal waters, or of

proceeding to internal waters, or of making for the high seas from internal waters.

3. Passage includes stopping and anchoring, but only in so far as the same are incidental to ordinary navigation or are rendered necessary by *force majeure* or by distress.

4. Passage is innocent so long as it is not prejudicial to the peace, good order or security of the coastal State. Such passage shall take place in conformity with these articles and with other rules of international law.

5. Passage of foreign fishing vessels shall not be considered innocent if they do not observe such laws and regulations as the coastal State may make and publish in order to prevent these vessels from fishing in the territorial sea.

6. Submarines are required to navigate on the surface and to show their flag.

Article 15

1. The coastal State must not hamper innocent passage through the territorial sea.

2. The coastal State is required to give appropriate publicity to any dangers to navigation, of which it has knowledge, within its territorial sea.

Article 16

1. The coastal State may take the necessary steps in its territorial sea to prevent passage which is not innocent.

2. In the case of ships proceeding to internal waters, the coastal State shall also have the right to take the necessary steps to prevent any breach of the conditions to which admission of those ships to those waters is subject.

3. Subject to the provisions of paragraph 4, the coastal State may, without discrimination amongst foreign ships, suspend temporarily in specified areas of its territorial sea the innocent passage of foreign ships if such suspension is essential for the protection of its security. Such suspension shall take effect only after having been duly published.

4. There shall be no suspension of the innocent passage of foreign ships through straits which are used for international navigation between one part of the high seas and another part of the high seas or the territorial sea of a foreign State.

Article 17

Foreign ships exercising the right of innocent passage shall comply with the laws and regulations enacted by the coastal State in conformity with these articles and other rules of international law and, in particular, with such laws and regulations relating to transport and navigation.

SUB-SECTION B. RULES APPLICABLE TO MERCHANT SHIPS

Article 18

1. No charge may be levied upon foreign ships by reason only of their passage through the territorial sea.

2. Charges may be levied upon a foreign ship passing through the territorial sea as payment only for specific services rendered to the ship. These charges shall be levied without discrimination.

Article 19

1. The criminal jurisdiction of the coastal State should not be exercised on board a foreign ship passing through the territorial sea to arrest any person or to conduct any investigation in connexion with any crime committed on board the ship during its passage, save only in the following cases:

(*a*) If the consequences of the crime extend to the coastal State; or

(*b*) If the crime is of a kind to disturb the peace of the country or the good order of the territorial sea; or

(*c*) If the assistance of the local authorities has been requested by the captain of the ship or by the consul of the country whose flag the ship flies; or

(*d*) If it is necessary for the suppression of illicit traffic in narcotic drugs.

2. The above provisions do not affect the right of the coastal State to take any steps authorized by its laws for the purpose of an arrest or investigation on board a foreign ship passing through the territorial sea after leaving internal waters.

3. In the cases provided for in paragraphs 1 and 2 of this article, the coastal State shall, if the captain so requests, advise the consular authority of the flag State before taking any steps, and shall facilitate contact between such authority and the ship's crew. In cases of emergency this notification may be communicated while the measures are being taken.

4. In considering whether or how an arrest should be made, the local authorities shall pay due regard to the interests of navigation.

5. The coastal State may not take any steps on board a foreign ship passing through the territorial sea to arrest any person or to conduct any investigation in connexion with any crime committed before the ship entered the territorial sea, if the ship, proceeding from a foreign port, is only passing through the territorial sea without entering internal waters.

Article 20

1. The coastal State should not stop or divert a foreign ship passing through the territorial sea for the purpose of exercising civil jurisdiction in relation to a person on board the ship.

2. The coastal State may not levy execution against or arrest the ship for the purpose of any civil proceedings, save only in respect of obligations or liabilities assumed or incurred by the ship itself in the course or for the purpose of its voyage through the waters of the coastal State.

3. The provisions of the previous paragraph are without prejudice to the right of the coastal State, in accordance with its laws, to levy execution against or to arrest, for the purpose of any civil proceedings, a foreign ship lying in the territorial sea, or passing through the territorial sea after leaving internal waters.

SUB-SECTION C. RULES APPLICABLE TO GOVERNMENT SHIPS . OTHER THAN WARSHIPS

Article 21

The rules contained in sub-sections A and B shall also apply to government ships operated for commercial purposes.

Article 22

1. The rules contained in sub-section A and in article 19 shall apply to government ships operated for non-commercial purposes.

2. With such exceptions as are contained in the provisions referred to in the preceding paragraph, nothing in these articles affects the immunities which such ships enjoy under these articles or other rules of international law.

SUB-SECTION D. RULE APPLICABLE TO WARSHIPS

Article 23

If any warship does not comply with the regulations of the coastal State concerning passage through the territorial sea and disregards any request for compliance which is made to it, the coastal State may require the warship to leave the territorial sea.

Part II: Contiguous Zone

Article 24

1. In a zone of the high seas contiguous to its territorial sea, the coastal State may exercise the control necessary to:

(*a*) Prevent infringement of its customs, fiscal, immigration or sanitary regulations within its territory or territorial sea;

(*b*) Punish infringement of the above regulations committed within its territory or territorial sea.

2. The contiguous zone may not extend beyond twelve miles from the baseline from which the breadth of the territorial sea is measured.

3. Where the coasts of two States are opposite or adjacent to each other, neither of the two States is entitled, failing agreement between them to the contrary, to extend its contiguous zone beyond the median line every point of which is equidistant from the nearest points on the baselines from which the breadth of the territorial seas of the two States is measured.

Part III: Final Articles

Article 25

The provisions of this Convention shall not affect conventions or other international agreements already in force, as between States Parties to them.

[Articles 26 to 32 inclusive are procedural in nature and have been omitted.]

B. CONVENTION ON THE CONTINENTAL SHELF*

The States Parties to this Convention, Have agreed as follows:

Article 1

For the purpose of these articles, the term "continental shelf" is used as referring (*a*) to the seabed and subsoil of the submarine areas adjacent to the coast but outside the area of the territorial sea, to a depth of 200 metres or, beyond that limit, to where the depth of the superjacent waters admits of the exploitation of the natural resources of the said areas; (*b*) to the seabed and subsoil of similar submarine areas adjacent to the coasts of islands.

Article 2

1. The coastal State exercises over the continental shelf sovereign rights for the purpose of exploring it and exploiting its natural resources.

2. The rights referred to in paragraph 1 of this article are exclusive in the sense that if the coastal State does not explore the continental shelf or exploit its natural resources, no one may undertake these activities, or make a claim to the continental shelf, without the express consent of the coastal State.

3. The rights of the coastal State over the continental shelf do not depend on occupation, effective or notional, or on any express proclamation.

* Adopted Apr. 26, 1958 (U.N. Doc. A/Conf. 13/L.55).

4. The natural resources referred to in these articles consist of the mineral and other non-living resources of the seabed and subsoil together with living organisms belonging to sedentary species, that is to say, organisms which, at the harvestable stage, either are immobile on or under the seabed or are unable to move except in constant physical contact with the seabed or the subsoil.

Article 3

The rights of the coastal State over the continental shelf do not affect the legal status of the superjacent waters as high seas, or that of the airspace above those waters.

Article 4

Subject to its right to take reasonable measures for the exploration of the continental shelf and the exploitation of its natural resources, the coastal State may not impede the laying or maintenance of submarine cables or pipelines on the continental shelf.

Article 5

1. The exploration of the continental shelf and the exploitation of its natural resources must not result in any unjustifiable interference with navigation, fishing or the conservation of the living resources of the sea, nor result in any interference with fundamental oceanographic or other scientific research carried out with the intention of open publication.

2. Subject to the provisions of paragraphs 1 and 6 of this article, the coastal State is entitled to construct and maintain or operate on the continental shelf installations and other devices necessary for its exploration and the exploitation of its natural resources, and to establish safety zones around such installations and devices and to take in those zones measures necessary for their protection.

3. The safety zones referred to in paragraph 2 of this article may extend to a distance of 500 metres around the installations and other devices which have been erected, measured from each point of their outer edge. Ships of all nationalities must respect these safety zones.

4. Such installations and devices, though under the jurisdiction of the coastal State, do not possess the status of islands. They have no territorial sea of their own, and their presence does not affect the delimitation of the territorial sea of the coastal State.

5. Due notice must be given of the construction of any such installations, and permanent means for giving warning of their presence must be maintained. Any installations which are abandoned or disused must be entirely removed.

6. Neither the installations or devices, nor the safety zones around

them, may be established where interference may be caused to the use of recognized sea lanes essential to international navigation.

7. The coastal State is obliged to undertake, in the safety zones, all appropriate measures for the protection of the living resources of the sea from harmful agents.

8. The consent of the coastal State shall be obtained in respect of any research concerning the continental shelf and undertaken there. Nevertheless, the coastal State shall not normally withhold its consent if the request is submitted by a qualified institution with a view to purely scientific research into the physical or biological characteristics of the continental shelf, subject to the proviso that the coastal State shall have the right, if it so desires, to participate or to be represented in the research, and that in any event the results shall be published.

Article 6

1. Where the same continental shelf is adjacent to the territories of two or more States whose coasts are opposite to each other, the boundary of the continental shelf appertaining to such States shall be determined by agreement between them. In the absence of agreement, and unless another boundary line is justified by special circumstances, the boundary is the median line, every point of which is equidistant from the nearest points of the baselines from which the breadth of the territorial sea of each State is measured.

2. Where the same continental shelf is adjacent to the territories of two adjacent States, the boundary of the continental shelf shall be determined by agreement between them. In the absence of agreement, and unless another boundary line is justified by special circumstances, the boundary shall be determined by application of the principle of equidistance from the nearest points of the baselines from which the breadth of the territorial sea of each State is measured.

3. In delimiting the boundaries of the continental shelf, any lines which are drawn in accordance with the principles set out in paragraphs 1 and 2 of this article should be defined with reference to charts and geographical features as they exist at a particular date, and reference should be made to fixed permanent identifiable points on the land.

Article 7

The provisions of these articles shall not prejudice the right of the coastal State to exploit the subsoil by means of tunnelling irrespective of the depth of water above the subsoil.

[Articles 8 to 15 inclusive are procedural in nature and have been omitted.]

C. CONVENTION ON THE HIGH SEAS*

The States Parties to this Convention,

Desiring to codify the rules of international law relating to the high seas,

Recognizing that the United Nations Conference on the Law of the Sea, held at Geneva from 24 February to 27 April 1958, adopted the following provisions as generally declaratory of established principles of international law,

Have agreed as follows:

Article 1

The term "high seas" means all parts of the sea that are not included in the territorial sea or in the internal waters of a State.

Article 2

The high seas being open to all nations, no State may validly purport to subject any part of them to its sovereignty. Freedom of the high seas is exercised under the conditions laid down by these articles and by the other rules of international law. It comprises, *inter alia,* both for coastal and non-coastal States:

1. Freedom of navigation;
2. Freedom of fishing;
3. Freedom to lay submarine cables and pipelines;
4. Freedom to fly over the high seas.

These freedoms, and others which are recognized by the general principles of international law, shall be exercised by all States with reasonable regard to the interests of other States in their exercise of the freedom of the high seas.

Article 3

1. In order to enjoy the freedom of the seas on equal terms with coastal States, States having no sea-coast should have free access to the sea. To this end States situated between the sea and a State having no sea-coast shall by common agreement with the latter and in conformity with existing international conventions accord:

   (*a*) To the State having no sea-coast, on a basis of reciprocity, free transit through their territory; and

   (*b*) To ships flying the flag of that State treatment equal to that accorded to their own ships, or to the ships of any other States, as regards access to seaports and the use of such ports.

---

* Adopted Apr. 26, 1958 (U.N. Doc. A/Conf. 13/L.53).

2. States situated between the sea and a State having no sea-coast shall settle, by mutual agreement with the latter, and taking into account the rights of the coastal State or State of transit and the special conditions of the State having no sea-coast, all matters relating to freedom of transit and equal treatment in ports, in case such States are not already parties to existing international conventions.

Article 4

Every State, whether coastal or not, has the right to sail ships under its flag on the high seas.

Article 5

1. Each State shall fix the conditions for the grant of its nationality to ships, for the registration of ships in its territory, and for the right to fly its flag. Ships have the nationality of the State whose flag they are entitled to fly. There must exist a genuine link between the State and the ship; in particular, the State must effectively exercise its jurisdiction and control in administrative, technical and social matters over ships flying its flag.

2. Each State shall issue to ships to which it has granted the right to fly its flag documents to that effect.

Article 6

1. Ships shall sail under the flag of one State only and, save in exceptional cases expressly provided for in international treaties or in these articles, shall be subject to its exclusive jurisdiction on the high seas. A ship may not change its flag during a voyage or while in a port of call, save in the case of a real transfer of ownership or change of registry.

2. A ship which sails under the flags of two or more States, using them according to convenience, may not claim any of the nationalities in question with respect to any other State, and may be assimilated to a ship without nationality.

Article 7

The provisions of the preceding articles do not prejudice the question of ships employed in the official service of an intergovernmental organization flying the flag of the organization.

Article 8

1. Warships on the high seas have complete immunity from the jurisdiction of any State other than the flag State.

2. For the purposes of these articles, the term "warship" means a ship belonging to the naval forces of a State and bearing the external marks

distinguishing warships of its nationality, under the command of an officer duly commissioned by the government and whose name appears in the Navy List, and manned by a crew who are under regular naval discipline.

Article 9

Ships owned or operated by a State and used only on government non-commercial service shall, on the high seas, have complete immunity from the jurisdiction of any State other than the flag State.

Article 10

1. Every State shall take such measures for ships under its flag as are necessary to ensure safety at sea with regard *inter alia* to:
   (*a*) The use of signals, the maintenance of communications and the prevention of collisions;
   (*b*) The manning of ships and labour conditions for crews taking into account the applicable international labor instruments;
   (*c*) The construction, equipment and seaworthiness of ships.

2. In taking such measures each State is required to conform to generally accepted international standards and to take any steps which may be necessary to ensure their observance.

Article 11

1. In the event of a collision or of any other incident of navigation concerning a ship on the high seas, involving the penal or disciplinary responsibility of the master or of any other person in the service of the ship, no penal or disciplinary proceedings may be instituted against such persons except before the judicial or administrative authorities either of the flag State or of the State of which such person is a national.

2. In disciplinary matters, the State which has issued a master's certificate or a certificate of competence or licence shall alone be competent, after due legal process, to pronounce the withdrawal of such certificates, even if the holder is not a national of the State which issued them.

3. No arrest or detention of the ship, even as a measure of investigation, shall be ordered by any authorities other than those of the flag State.

Article 12

1. Every State shall require the master of a ship under its flag, in so far as he can do so without serious danger to the ship, the crew or the passengers:
   (*a*) To render assistance to any person found at sea in danger of being lost;
   (*b*) To proceed with all possible speed to the rescue of persons in

distress if informed of their need of assistance, in so far as such action may reasonably be expected of him;

(c) After a collision, to render assistance to the other ship, her crew and her passengers and, where possible, to inform the other ship of the name of his own ship, her port of registry and the nearest port at which she will call.

2. Every coastal State shall promote the establishment and maintenance of an adequate and effective search and rescue service regarding safety on and over the sea and—where circumstances so require—by way of mutual regional arrangements cooperate with neighbouring States for this purpose.

Article 13

Every State shall adopt effective measures to prevent and punish the transport of slaves in ships authorized to fly its flag, and to prevent the unlawful r ʌe of its flag for that purpose. Any slave taking refuge on board any sh·ᵖ, whatever its flag, shall, *ipso facto,* be free.

Article 14

All States shall co-operate to the fullest possible extent in the repression of piracy on the high seas or in any other place outside the jurisdiction of any State.

Article 15

Piracy consists of any of the following acts:

(1) Any illegal acts of violence, detention or any act of depredation, committed for private ends by the crew or the passengers of a private ship or a private aircraft, and directed:

(a) On the high seas, against another ship or aircraft, or against persons or property on board such ship or aircraft;

(b) Against a ship, aircraft, persons or property in a place outside the jurisdiction of any State;

(2) Any act of voluntary participation in the operation of a ship or of an aircraft with knowledge of facts making it a pirate ship or aircraft;

(3) Any act of inciting or of intentionally facilitating an act described in sub-paragraph 1 or sub-paragraph 2 of this article.

Article 16

The acts of piracy, as defined in article 15, committed by a warship, government ship or government aircraft whose crew has mutinied and taken control of the ship or aircraft are assimilated to acts committed by a private ship.

Article 17

A ship or aircraft is considered a pirate ship or aircraft if it is intended by the persons in dominant control to be used for the purpose of committing one of the acts referred to in article 15. The same applies if the ship or aircraft has been used to commit any such act, so long as it remains under the control of the persons guilty of that act.

Article 18

A ship or aircraft may retain its nationality although it has become a pirate ship or aircraft. The retention or loss of nationality is determined by the law of the State from which such nationality was derived.

Article 19

On the high seas, or in any other place outside the jurisdiction of any State, every State may seize a pirate ship or aircraft, or a ship taken by piracy and under the control of pirates, and arrest the persons and seize the property on board. The courts of the State which carried out the seizure may decide upon the penalties to be imposed, and may also determine the action to be taken with regard to the ships, aircraft or property, subject to the rights of third parties acting in good faith.

Article 20

Where the seizure of a ship or aircraft on suspicion of piracy has been effected without adequate grounds, the State making the seizure shall be liable to the State the nationality of which is possessed by the ship or aircraft, for any loss or damage caused by the seizure.

Article 21

A seizure on account of piracy may only be carried out by warships or military aircraft, or other ships or aircraft on government service authorized to that effect.

Article 22

1. Except where acts of interference derive from powers conferred by treaty, a warship which encounters a foreign merchant ship on the high seas is not justified in boarding her unless there is reasonable ground for suspecting:

(*a*) That the ship is engaged in piracy; or

(*b*) That the ship is engaged in the slave trade; or

(*c*) That, though flying a foreign flag or refusing to show its flag, the ship is, in reality, of the same nationality as the warship.

2. In the cases provided for in sub-paragraphs (*a*), (*b*) and (*c*)

above, the warship may proceed to verify the ship's right to fly its flag. To this end, it may send a boat under the command of an officer to the suspected ship. If suspicion remains after the documents have been checked, it may proceed to a further examination on board the ship, which must be carried out with all possible consideration.

3. If the suspicions prove to be unfounded, and provided that the ship boarded has not committed any act justifying them, it shall be compensated for any loss or damage that may have been sustained.

Article 23

1. The hot pursuit of a foreign ship may be undertaken when the competent authorities of the coastal State have good reason to believe that the ship has violated the laws and regulations of that State. Such pursuit must be commenced when the foreign ship or one of its boats is within the internal waters or the territorial sea or the contiguous zone of the pursuing State, and may only be continued outside the territorial sea or the contiguous zone if the pursuit has not been interrupted. It is not necessary that, at the time when the foreign ship within the territorial sea or the contiguous zone receives the order to stop, the ship giving the order should likewise be within the territorial sea or the contiguous zone. If the foreign ship is within a contiguous zone, as defined in article 24 of the Convention on the Territorial Sea and the Contiguous Zone, the pursuit may only be undertaken if there has been a violation of the rights for the protection of which the zone was established.

2. The right of hot pursuit ceases as soon as the ship pursued enters the territorial sea of its own country or of a third State.

3. Hot pursuit is not deemed to have begun unless the pursuing ship has satisfied itself by such practicable means as may be available that the ship pursued or one of its boats or other craft working as a team and using the ship pursued as a mother ship are within the limits of the territorial sea, or as the case may be within the contiguous zone. The pursuit may only be commenced after a visual or auditory signal to stop has been given at a distance which enables it to be seen or heard by the foreign ship.

4. The right of hot pursuit may be exercised only by warships or military aircraft, or other ships or aircraft on government service specially authorized to that effect.

5. Where hot pursuit is effected by an aircraft:

   (*a*) The provisions of paragraphs 1 to 3 of this article shall apply *mutatis mutandis;*
   (*b*) The aircraft giving the order to stop must itself actively pursue the ship until a ship or aircraft of the coastal State, summoned by the aircraft, arrives to take over the pursuit, unless the aircraft is

itself able to arrest the ship. It does not suffice to justify an arrest on the high seas that the ship was merely sighted by the aircraft as an offender or suspected offender, if it was not both ordered to stop and pursued by the aircraft itself or other aircraft or ships which continue the pursuit without interruption.

6. The release of a ship arrested within the jurisdiction of a State and escorted to a port of that State for the purposes of an inquiry before the competent authorities, may not be claimed solely on the ground that the ship, in the course of its voyage, was escorted across a portion of the high seas, if the circumstances rendered this necessary.

7. Where a ship has been stopped or arrested on the high seas in circumstances which do not justify the exercise of the right of hot pursuit, it shall be compensated for any loss or damage that may have been thereby sustained.

Article 24

Every State shall draw up regulations to prevent pollution of the seas by the discharge of oil from ships or pipelines or resulting from the exploitation and exploration of the seabed and its subsoil, taking account of existing treaty provisions on the subject.

Article 25

1. Every State shall take measures to prevent pollution of the seas from the dumping of radioactive waste, taking into account any standards and regulations which may be formulated by the competent international organizations.

2. All States shall co-operate with the competent international organizations in taking measures for the prevention of pollution of the seas or air space above, resulting from any activities with radioactive materials or other harmful agents.

Article 26

1. All States shall be entitled to lay submarine cables and pipelines on the bed of the high seas.

2. Subject to its right to take reasonable measures for the exploration of the continental shelf and the exploitation of its natural resources, the coastal State may not impede the laying or maintenance of such cables or pipelines.

3. When laying such cables or pipelines the State in question shall pay due regard to cables or pipelines already in position on the seabed. In particular, possibilities of repairing existing cables or pipelines shall not be prejudiced.

Article 27

Every State shall take the necessary legislative measures to provide that the breaking or injury by a ship flying its flag or by a person subject to its jurisdiction of a submarine cable beneath the high seas done wilfully or through culpable negligence, in such a manner as to be liable to interrupt or obstruct telegraphic or telephonic communications, and similarly the breaking or injury of a submarine pipeline or high-voltage power cable shall be a punishable offense. This provision shall not apply to any break or injury caused by persons who acted merely with the legitimate object of saving their lives or their ships, after having taken all necessary precautions to avoid such break or injury.

Article 28

Every State shall take the necessary legislative measures to provide that, if persons subject to its jurisdiction who are the owners of a cable or pipeline beneath the high seas, in laying or repairing that cable or pipeline, cause a break in or injury to another cable or pipeline, they shall bear the cost of the repairs.

Article 29

Every State shall take the necessary legislative measures to ensure that the owners of ships who can prove that they have sacrificed an anchor, a net or any other fishing gear, in order to avoid injuring a submarine cable or pipeline, shall be indemnified by the owner of the cable or pipeline, provided that the owner of the ship has taken all reasonable precautionary measures beforehand.

Article 30

The provisions of this Convention shall not affect conventions or other international agreements already in force, as between States parties to them.

[Articles 31 to 37 inclusive are procedural in nature and have been omitted.]

D. CONVENTION ON FISHING AND CONSERVATION OF THE LIVING RESOURCES OF THE HIGH SEAS*

The States Parties to this Convention,
Considering that the development of modern techniques for the exploitation of the living resources of the sea, increasing man's ability to meet

---

* Adopted Apr. 26, 1958 (U.N. Doc. A/Conf. 13/L.54).

the need of the world's expanding population for food, has exposed some of these resources to the danger of being over-exploited,

Considering also that the nature of the problems involved in the conservation of the living resources of the high seas is such that there is a clear necessity that they be solved, whenever possible, on the basis of international co-operation through the concerted action of all the States concerned,

Have agreed as follows:

Article 1

1. All States have the right for their nationals to engage in fishing on the high seas, subject (*a*) to their treaty obligations, (*b*) to the interests and rights of coastal States as provided for in this Convention, and (*c*) to the provisions contained in the following articles concerning conservation of the living resources of the high seas.

2. All States have their duty to adopt, or to co-operate with other States in adopting, such measures for their respective nationals as may be necessary for the conservation of the living resources of the high seas.

Article 2

As employed in this Convention, the expression "conservation of the living resources of the high seas" means the aggregate of the measures rendering possible the optimum sustainable yield from those resources so as to secure a maximum supply of food and other marine products. Conservation programmes should be formulated with a view to securing in the first place a supply of food for human consumption.

Article 3

A State whose nationals are engaged in fishing any stock or stocks of fish or other living marine resources in any area of the high seas where the nationals of other States are not thus engaged shall adopt, for its own nationals, measures in that area when necessary for the purpose of the conservation of the living resources affected.

Article 4

1. If the nationals of two or more States are engaged in fishing the same stock or stocks of fish or other living marine resources in any area or areas of the high seas, these States shall, at the request of any of them, enter into negotiations with a view to prescribing by agreement for their nationals the necessary measures for the conservation of the living resources affected.

2. If the States concerned do not reach agreement within twelve months, any of the parties may initiate the procedure contemplated by article 9.

Article 5

1. If, subsequent to the adoption of the measures referred to in articles 3 and 4, nationals of other States engage in fishing the same stock or stocks of fish or other living marine resources in any area or areas of the high seas, the other States shall apply the measures, which shall not be discriminatory in form or in fact, to their own nationals not later than seven months after the date on which the measures shall have been notified to the Director-General of the Food and Agriculture Organization of the United Nations. The Director-General shall notify such measures to any State which so requests and, in any case, to any State specified by the State initiating the measure.

2. If these other States do not accept the measures so adopted and if no agreement can be reached within twelve months, any of the interested parties may initiate the procedure contemplated by article 9. Subject to paragraph 2 of article 10, the measures adopted shall remain obligatory pending the decision of the special commission.

Article 6

1. A coastal State has a special interest in the maintenance of the productivity of the living resources in any area of the high seas adjacent to its territorial sea.

2. A coastal State is entitled to take part on an equal footing in any system of research and regulation for purposes of conservation of the living resources of the high seas in that area, even though its nationals do not carry on fishing there.

3. A State whose nationals are engaged in fishing in any area of the high seas adjacent to the territorial sea of a coastal State shall, at the request of that coastal State, enter into negotiations with a view to prescribing by agreement the measures necessary for the conservation of the living resources of the high seas in that area.

4. A State whose nationals are engaged in fishing in any area of the high seas adjacent to the territorial sea of a coastal State shall not enforce conservation measures in that area which are opposed to those which have been adopted by the coastal State, but may enter into negotiations with the coastal State with a view to prescribing by agreement the measures necessary for the conservation of the living resources of the high seas in that area.

5. If the States concerned do not reach agreement with respect to conservation measures within twelve months, any of the parties may initiate the procedure contemplated by article 9.

Article 7

1. Having regard to the provisions of paragraph 1 of article 6, any coastal State may, with a view to the maintenance of the productivity of the

living resources of the sea, adopt unilateral measures of conservation appropriate to any stock of fish or other marine resources in any area of the high seas adjacent to its territorial sea, provided that negotiations to that effect with the other States concerned have not led to an agreement within six months.

2. The measures which the coastal State adopts under the previous paragraph shall be valid as to other States only if the following requirements are fulfilled:

(*a*) That there is a need for urgent application of conservation measures in the light of the existing knowledge of the fishery;

(*b*) That the measures adopted are based on appropriate scientific findings;

(*c*) That such measures do not discriminate in form or in fact against foreign fishermen.

3. These measures shall remain in force pending the settlement, in accordance with the relevant provisions of this Convention, of any disagreement as to their validity.

4. If the measures are not accepted by the other States concerned, any of the parties may initiate the procedure contemplated by article 9. Subject to paragraph 2 of article 10, the measures adopted shall remain obligatory pending the decision of the special commission.

5. The principles of geographical demarcation as defined in article 12 of the Convention on the Territorial Sea and the Contiguous Zone shall be adopted when coasts of different States are involved.

Article 8

1. Any State which, even if its nationals are not engaged in fishing in an area of the high seas not adjacent to its coast, has a special interest in the conservation of the living resources of the high seas in that area, may request the State or States whose nationals are engaged in fishing there to take the necessary measures of conservation under articles 3 and 4 respectively, at the same time mentioning the scientific reasons which in its opinion make such measures necessary, and indicating its special interest.

2. If no agreement is reached within twelve months, such State may initiate the procedure contemplated by article 9.

Article 9

1. Any dispute which may arise between States under articles 4, 5, 6, 7 and 8 shall, at the request of any of the parties, be submitted for settlement to a special commission of five members, unless the parties agree to seek a solution by another method of peaceful settlement, as provided for in Article 33 of the Charter of the United Nations.

2. The members of the commission, one of whom shall be designated

as chairman, shall be named by agreement between the States in dispute within three months of the request for settlement in accordance with the provisions of this article. Failing agreement they shall, upon the request of any State party, be named by the Secretary-General of the United Nations, within a further three-month period, in consultation with the States in dispute and with the President of the International Court of Justice and the Director-General of the Food and Agricultural Organization of the United Nations, from amongst well-qualified persons being nationals of States not involved in the dispute and specializing in legal, administrative or scientific questions relating to fisheries, depending upon the nature of the dispute to be settled. Any vacancy arising after the original appointment shall be filled in the same manner as provided for the initial selection.

3. Any State party to proceedings under these articles shall have the right to name one of its nationals to the special commission, with the right to participate fully in the proceedings on the same footing as a member of the commission, but without the right to vote or to take part in the writing of the commission's decision.

4. The commission shall determine its own procedure, assuring each party to the proceedings a full opportunity to be heard and to present its case. It shall also determine how the costs and expenses shall be divided between the parties to the dispute, failing agreement by the parties on this matter.

5. The special commission shall render its decision within a period of five months from the time it is appointed unless it decides, in case of necessity, to extend the time limit for a period not exceeding three months.

6. The special commission shall, in reaching its decisions, adhere to these articles and to any special agreements between the disputing parties regarding settlement of the dispute.

7. Decisions of the commission shall be by majority vote.

Article 10

1. The special commission shall, in disputes arising under article 7, apply the criteria listed in paragraph 2 of that article. In disputes under articles 4, 5, 6 and 8, the commission shall apply the following criteria, according to the issues involved in the dispute:

(a) Common to the determination of disputes arising under articles 4, 5 and 6 are the requirements:

(i) That scientific findings demonstrate the necessity of conservation measures;

(ii) That the specific measures are based on scientific findings and are practicable; and

(iii) That the measures do not discriminate, in form or in fact, against fishermen of other States.

(*b*) Applicable to the determination of disputes arising under article 8 is the requirement that scientific findings demonstrate the necessity for conservation measures, or that the conservation programme is adequate, as the case may be.

2. The special commission may decide that pending its award the measures in dispute shall not be applied, provided that, in the case of disputes under article 7, the measures shall only be suspended when it is apparent to the commission on the basis of *prima facie* evidence that the need for the urgent application of such measures does not exist.

Article 11

The decisions of the special commission shall be binding on the States concerned and the provisions of paragraph 2 of Article 94 of the Charter of the United Nations shall be applicable to those decisions. If the decisions are accompanied by any recommendations, they shall receive the greatest possible consideration.

Article 12

1. If the factual basis of the award of the special commission is altered by substantial changes in the conditions of the stock or stocks of fish or other living marine resources or in methods of fishing, any of the States concerned may request the other States to enter into negotiations with a view to prescribing by agreement the necessary modifications in the measures of conservation.

2. If no agreement is reached within a reasonable period of time, any of the States concerned may again resort to the procedure contemplated by article 9 provided that at least two years have elapsed from the original award.

Article 13

1. The regulation of fisheries conducted by means of equipment embedded in the floor of the sea in areas of the high seas adjacent to the territorial sea of a State may be undertaken by that State where such fisheries have long been maintained and conducted by its nationals, provided that non-nationals are permitted to participate in such activities on an equal footing with nationals except in areas where such fisheries have by long usage been exclusively enjoyed by such nationals. Such regulations will not, however, affect the general status of the areas as high seas.

2. In this article, the expression "fisheries conducted by means of equipment embedded in the floor of the sea" means those fisheries using gear with supporting members embedded in the sea floor, constructed on a site and left there to operate permanently or, if removed, restored each season on the same site.

Article 14

In articles 1, 3, 4, 5, 6 and 8, the term "nationals" means fishing boats or craft of any size having the nationality of the State concerned, according to the law of that State, irrespective of the nationality of the members of their crews.

[Articles 15 to 22 inclusive are procedural in nature and have been omitted.]

E. OPTIONAL PROTOCOL OF SIGNATURE CONCERNING THE COMPULSORY SETTLEMENT OF DISPUTES*

The States parties to this Protocol and to any one or more of the Conventions on the Law of the Sea adopted by the United Nations Conference on the Law of the Sea held at Geneva from 24 February to 27 April 1958.

Expressing their wish to resort, in all matters concerning them in respect of any dispute arising out of the interpretation or application of any article of any Convention on the Law of the Sea of 29 April 1958, to the compulsory jurisdiction of the International Court of Justice, unless some other form of settlement is provided in the Convention or has been agreed upon by the parties within a reasonable period,

Have agreed as follows:

Article I

Disputes arising out of the interpretation or application of any Convention on the Law of the Sea shall lie within the compulsory jurisdiction of the International Court of Justice, and may accordingly be brought before the Court by an application made by any party to the dispute being a party to this Protocol.

Article II

This undertaking relates to all the provisions of any Convention on the Law of the Sea except, in the Convention on Fishing and Conservation of the Living Resources of the High Seas, articles 4, 5, 6, 7 and 8, to which articles 9, 10, 11 and 12 of that Convention remain applicable.

Article III

The parties may agree, within a period of two months after one party has notified its opinion to the other that a dispute exists, to resort not to the International Court of Justice but to an arbitral tribunal. After the expiry of the said period, either party to this Protocol may bring the dispute before the Court by an application.

* Adopted Apr. 26, 1958 (U.N. Doc. A/Conf. 13/L.57).

Article IV

1. Within the same period of two months, the parties to this Protocol may agree to adopt a conciliation procedure before resorting to the International Court of Justice. "

2. The conciliation commission shall make its recommendations within five months after its appointment. If its recommendations are not accepted by the parties to the dispute within two months after they have been delivered, either party may bring the dispute before the Court by an application.

[Articles V to VII inclusive are procedural in nature and have been omitted.]

Appendix 10

Cost Indices of Petroleum Exploration Drilling, Development Drilling, Production Facilities, and Pipelines*

A. EXPLORATION DRILLING

Figure A 10.1 illustrates the variation of total daily drilling cost with water depth for the Gulf of Mexico. Mobile rigs are generally used for exploration purposes when dry hole risk is at a maximum. As shown in *Figure A 10.1,* the daily drilling cost from mobile rigs in 600 feet (183 meters) of water is 80 percent greater than in 100 feet (30 meters) of water, and twice as great in 1,000 feet (305 meters) of water as in 100 feet of water. It is assumed that the cost of exploration drilling is directly proportional to these daily drilling costs.

It is currently thought that present floating drilling techniques can be extended to water depths of at least 1,500 feet (457 meters).[1] Beyond 1,500 feet, several changes in the existing mobile drilling techniques will be necessary. Three of these changes are described below, and each of them will significantly increase the drilling cost trend.

1. Dynamic Positioning

With present marine conductor drilling systems the lateral movements of a floating drilling unit must be kept within 5 or 10 percent of the water depth to permit operations. Most units have eight or more 20,000- to 30,000-lb. anchors. These systems are heavy, costly, and cumbersome, and

* From: *Petroleum Resources under the Ocean Floor, March 1969.* Washington, National Petroleum Council, 1969. Note.—The cost data presented in this appendix are rough approximations based on the limited amount of published data available at the time of writing this report. While isolated values may prove to be in error as more data becomes available, it is felt that the overall conclusions of this analysis as presented in the main text of the report will remain valid.

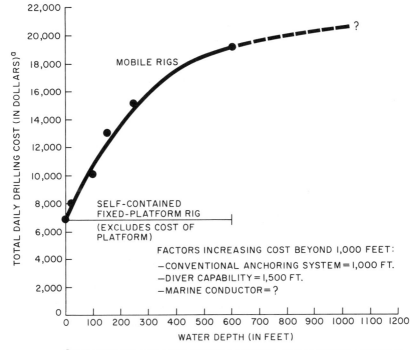

Figure A 10.1. Total daily offshore drilling cost—Gulf of Mexico.

are not believed to be feasible in open ocean in water depths beyond about 1,000–1,500 feet (305–457 meters).[2]

For this reason, dynamic positioning systems have been developed and employed on coring vessels and several ship-shaped drilling vessels. These systems consist of a position indicator, an analog computer, and a propulsion system capable of producing thrusts of a variable direction and magnitude on command from the computer. There are currently four large ship-shape drilling vessels under construction which will use dynamic positioning. The largest is the 50,000-ton *Noess Crusader,* whose computer and indicating system alone will cost in excess of one million dollars.[3]

Thus, the capital costs of dynamic positioning systems are high, and because an exploratory well can take 100 days to drill and evaluate, operating and maintenance costs could be significant. That dynamic positioning is costly is illustrated by the fact that at least three of the four drill vessels mentioned above will be equipped with conventional anchor systems for use in most drilling operations.[4]

2. Diver Capability

The deepest working dive on record was to a depth of 700 feet (213 meters) and consisted of a 15-minute biological survey.[5] Laboratory dives have been made to depths slightly greater than 1,000 feet (305 meters),[6] but it appears as if diver capability will be limited, for the foreseeable future, to less than 1,500 feet (457 meters). While drilling and completion systems have been developed to eliminate the need for divers, it has been found that even with these systems divers can be economically utilized to handle unexpected occurrences. The use of submersibles with attached manipulators in depths greater than the limits of diver capability will be costly.

3. Marine Conductor

The marine conductor extends from the wellhead, which is fixed on the ocean floor, to the floating drilling vessel. This conductor guides the drill pipe and bit down to the underwater wellhead and into the well and provides a return path for the drilling fluid and cuttings. The guide function insures that as bits wear out and must be replaced the drill string will be able to re-enter the hole; and the mud returns are desirable to decrease mud costs, aid in controlling the well, and also give an evaluation of drilling progress.

Because marine conductors designed to withstand wave and current forces in deep water are costly, they are not used in deep water, shallow depth core drilling programs. The core hole is drilled until the bit is worn out and the hole is then abandoned. A re-entry drilling system without a marine conductor has not been developed.

As water depth increases, the cost of a marine conductor also will increase, until at some point a new system for drilling wells may have to be developed.

B. DEVELOPMENT DRILLING

1. Platforms

Generally, when a large field is discovered, a fixed platform is installed so that development wells can be drilled with a lower cost platform rig. The platform may also support some production facilities. The exploratory wells are either completed underwater and produced to the platform, or they may be abandoned. Present platform technology indicates the industry can develop the capability to erect platforms in 600 feet (183 meters) of water,[7] but beyond this depth, all development wells may have to be drilled with mobile units and completed on the ocean floor.

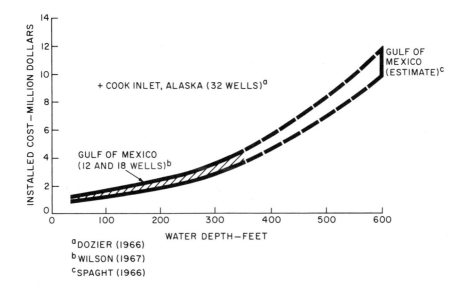

Figure A 10.2. Cost data and estimates for self-contained platforms.

Figure A 10.2 shows platform costs can be expected to increase sharply as the water depth increases. The cost for existing platforms in the Gulf of Mexico varies from $1.5 million in 100 feet (30 meters) of water to $4 million in 350 feet (107 meters) of water. The cost of platforms in 600 feet (183 meters) of water is expected to approach $12 million, or six to eight times as costly as those in 100 feet of water. The graph also indicates the effect of environmental conditions on platform costs, with platforms in Alaska's Cook Inlet costing about six times as much as Gulf of Mexico platforms for the same water depth.

There are locations, such as in the Santa Barbara Channel of California, where the seabed has a steep slope and where, even though a development prospect may be in over 600 feet (183 meters) of water, it is possible to exploit it with directional drilling from a platform in 600 feet of water. Nineteen of the seventy-one leases purchased in the February 1968 Santa Barbara sale are in water depths too great to support a 600-foot structure on the lease. These leases, which represent about 10 percent of the total bonus paid, lie entirely within 5 miles of the 600-foot contour, and a number of operators are undoubtedly planning on developing many of the tracts by directional drilling from platforms in 600 feet of water or less, or by completing wells under water and producing them to platforms in shallower water.[8]

2. Underwater Completions

Over 100 wells have been completed to date using underwater completion techniques.[9] The techniques are well known, and the equipment has been proven. Because of the necessity of providing for well re-entry to repair downhole equipment, remove paraffin, and to perform sand consolidation and various other remedial work, costs of underwater completions are very high.

In the Gulf of Mexico it costs about $550,000 to drill and complete a 12,000-foot underwater exploratory well in 100 feet (30 meters) of water and to install a flow line to a nearby platform.[10] The cost for underwater wells should increase in proportion to the total drilling cost of mobile rigs (*Figure A 10.1*) and should thus approach $990,000 in 600 feet (183 meters) of water, and $1,100,000 in 1,000 feet (305 meters) of water.

This can be compared to the $400,000 cost of drilling and completing a platform well in 150 feet (46 meters) of water, including a share of platform costs.[11] The cost of platform wells increases with water depth, as their share of the platform costs increase. From *Figure A 10.2,* the per well share of platform cost is approximately $80,000 in 100 feet (30 meters) of water, $100,000 in 150 feet of water, and $600,000 in 600 feet (183 meters) of water. Thus, the cost of a platform well can be estimated as $400,000 in 100 feet of water and $900,000 in 600 feet of water.

The latter figure is roughly equivalent to the cost of an underwater completion in this water depth. By projecting the curve of *Figure A 10.2* it appears that the initial costs of underwater development wells will be cheaper than platform wells in 1,000 feet (305 meters) of water. However, operating costs of underwater wells are relatively unknown at this time.

C. PRODUCTION FACILITIES

After the wells are drilled, production facilities are required to permit processing, measuring, storing, and transporting the crude oil to a sale point.

Offshore platform-mounted production facilities do not differ significantly from those onshore. They are more expensive, however, because of the need to minimize weight and space, and the higher costs of offshore construction and operation. Also, in most cases living quarters and support facilities must be provided, which further increase costs. Within platform capabilities, the cost for facilities should not increase significantly as water depth increases. In water depths beyond platform capability, wells will be completed under water and produced to an underwater or floating production facility which can be expected to cost significantly more than platform-mounted facilities.

Although several fields have been developed with underwater completions and flow lines from the individual wellheads to production facilities onshore, no underwater or floating production facilities have yet been installed. However, because of the demands placed on floating and underwater production equipment by the environment, as well as by the varied tasks which must be performed, it is clear that these facilities will be complex and costly.

In addition, because a diver or submersible vehicle is required for surveillance and maintenance, operating costs for underwater facilities will undoubtedly be much greater than those of existing surface facilities. Existing submersibles are mainly research vehicles; they do not presently have manipulators capable of meeting the torque requirements of the offshore oil industry,[12] and they are costly to build and operate.

D. PIPELINES

Offshore pipelines in the Gulf of Mexico are between two and 4½ times more costly than comparable onshore pipelines.[13] It is not possible to draw a graph of cost versus water depths from existing experience since costs depend also on location, size, weight and coating required, soil conditions, and wave and tide characteristics. For example, in Cook Inlet, Alaska, where 30-foot tides cause surface currents of 13 feet per second and bottom currents of 6 feet per second, and where severe weather conditions limit the laying season, the cost of an 8-inch line in 120 feet (37 meters) of water is approximately six times what it is in the Gulf of Mexico.[14]

1. Weather

The largest lay barges cannot operate in wave heights exceeding 10 to 15 feet and they experience difficulty transferring pipe in wave heights less than this.[15] Thus, as operations expand into deeper, more exposed locations, delays due to weather will increase and costs will rise.

2. Location

As pipelines are laid in more remote locations and further from shore, mobilization and transportation expenses will increase. For example, a lay barge and its support equipment may cost $20,000 per day in the Gulf of Mexico and $60,000 per day in Cook Inlet.[16]

3. Method of Pipe Laying

In water depths of less than 100 feet (30 meters), most sizes of pipe are self-supporting and can be laid without the need of elaborate guides or tensioning systems to decrease the curvature. Stingers have been used in water depths of up to 300 feet (91 meters) to guide the pipe from the lay barge to the bottom. However, in this water depth a 700-foot stinger is required, which is costly and cumbersome, and operations are plagued with continued damage or trouble with the stinger.[17]

To eliminate the need for a long stinger, tensioning devices are used in conjunction with small stingers in water depths greater than 250 feet (76 meters) and up to 600 feet (183 meters).[18] The tension required to maintain minimum curvature increases with size, and thus only small diameter pipelines can be laid with this method in water depths from approximately 600 to 1,000 feet (183 to 305 meters).[19] Beyond this depth new techniques will have to be employed.[20]

4. Pipeline Burying

In deep water, wave action is not as strongly felt on the bottom, and in areas where currents are not a problem, pipelines will not have to be buried. Since most operators are currently burying pipelines in water depths less than 100 to 200 feet (30 to 61 meters), the elimination of the need to bury will partially offset the trends discussed above.[21]

REFERENCES

1. "Humble Gearing Up to Drill in 1300 Feet," *Oil and Gas Journal,* May 6, 1968; "Offshore Report," *Oil and Gas Journal,* July 10, 1967.
2. J. R. Dozier, "Offshore Oil and Gas Operations, Present and Future," presented at Law of the Sea Institute, University of Rhode Island, June 1966.
3. "Where Dynamic Positioning Stands Now," *Offshore,* May 1968.
4. *Ibid.*
5. *Oil and Gas Journal,* April 22, 1968.
6. "Men Under Pressure," *Ocean Industry,* April 1968.
7. See footnote 1.
8. "Santa Barbara Channel—Promises and Problems," *Ocean Industry,* May 1968.
9. M. E. Spaght, "The Development of Underwater Oil and Gas Reserves," presented at Royal Swedish Acad. of Engineering Science, 1966.
10. "Offshore Report," *Oil and Gas Journal,* June 20, 1966.
11. *Ibid.*
12. *Resources of the Sea,* report of the Secretary General, United Nations Economic and Social Council, February 1968.
13. "Offshore Pipelining," *Oil and Gas Journal,* December 12, 1966.
14. M. E. Spaght, *loc. cit.*
15. "Offshore Pipelining," *loc. cit.*
16. *Ibid.*

17. J. R. Dozier, *loc. cit.*

18. H. M. Wilkinson and J. P. Fraser, "Deep Water Pipeline," paper (API Division of Production).

19. "Offshore Pipelining," *loc. cit.*

20. J. Delarvell, "The French Install Submarine Pipeline at Depth of 1080 Ft.," *Oil and Gas Journal,* August 7, 1967.

21. R. Blumber, "Hurricane Winds, Waves and Currents Test Marine Pipe Line Design," *Pipe Line Industry,* June–November 1964.

Appendix 11

Petroleum Operations in the Sea—
1980 and Beyond*

DR. RICHARD J. HOWE
ESSO PRODUCTION RESEARCH CO., HOUSTON

SUMMARY

During the past 22 years, the petroleum industry has developed the capability to drill for and produce oil and gas from the Continental shelves of the world. Exploratory wells have been drilled in 640 ft. of water and huge production platforms stand in water 340 ft. deep. By year-end, wells will be drilled in the Santa Barbara Channel in water depths ranging up to 1,300 ft. If these wells discover substantial reserves, suitable production facilities will be built to bring the oil and gas to market.

In a development which is accelerating so rapidly, it is indeed difficult to predict the status of offshore petroleum operations in 1980. It appears, however, that the water depths in which the petroleum industry will be operating in that year will be governed by economics rather than technical factors. By that time the industry should have the capability to explore for and produce hydrocarbon reserves in almost any ocean area of the world; however, there are many alternate sources of energy which will probably enter the market before petroleum deposits are produced in ultradeep water.

It is possible to extrapolate some current trends to 1980. It is quite clear that the use of floating drilling rigs will continue to expand as the industry moves into deeper water. Of the 350 mobile units operating in 1980, approximately 60 percent of them will be floaters compared with 35 percent today. There is also a possibility that a small number of undersea or remotely controlled drilling rigs will be in operation in 1980. An increasing number of wells will be installed on the sea floor. Many deepwater fields will be located completely below water. As equipment for these fields is perfected, it will also be used in shallower water. In 1980, oil production

* From: *Ocean Industry,* Gulf Publishing Co., Houston, Texas, August 1968, pp. 25–29.

from offshore fields will total 20 million barrels of oil per day, representing approximately one-third of total Free World production, compared with 5 million barrels and 17 percent today. Total investment in offshore petroleum resources will then stand at $55 billion, compared with $18 billion today.

Before discussing the future of offshore petroleum operations, it would be appropriate to review what has happened in the past and where we stand today.

Several years ago, Lewis Weeks, a leading petroleum consultant, made a comprehensive study to determine which areas of the world might be considered prospective for the accumulation of oil and gas. He estimated that of the 57 million square miles of land surface, approximately 18 might contain oil and gas deposits. Of the 140 million square miles of water-covered area, he predicts that there are 6 million square miles of favorable basin areas out of a total of 10 million square miles out to the 1,000-ft. contour. Thus, 25 percent of the world's prospective basin areas lie offshore in water depths less than 1,000 ft. (See Figure A 11.1.)

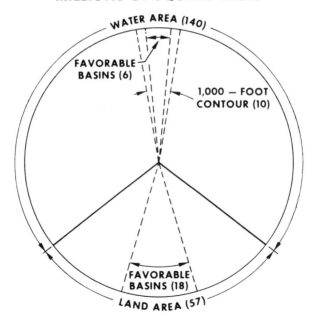

TOTAL AREA = 197 MILLION SQ. MILES

Figure A 11.1. Earth surface area.

Using this information, plus a number of other important geological factors such as the age and thickness of the sediments, Weeks predicted that the potential world reserves of petroleum are at least 2.2 trillion barrels of oil and natural gas liquids. These reserve figures apply to primary recovery. Supplemental recovery operations could increase the reserve figures by 50 percent or more. As indicated in Figure A 11.2, approximately one-third of the world's potential reserves lie offshore, yet only a small percentage of these reserves have been discovered to date. This is certainly an important factor behind the rapid expansion of offshore petroleum exploration and production.

With these incentives in mind, let's briefly review the major offshore drilling developments that have occurred during this century:

The first offshore drilling was carried out from wooden piers along the coast of California at the turn of the century. This was a natural extension of land fields into the sea.

During the 1920s the same thing occurred in Lake Maracaibo, Venezuela. More than 5,000 wells have now been drilled in the lake in water depths up to 120 ft.

During the next decade, there was extensive drilling in the marshes of Louisiana. Several hundred drilling rigs designed for this purpose were built during a 20-year period.

In the 1940s, drilling moved into the open water of the Gulf of Mexico. By 1948, platforms were being built in 50 ft. of water off Louisiana.

BILLION BARRELS OF OIL

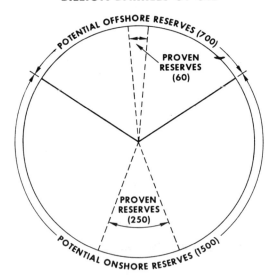

TOTAL EST. PRIMARY RESERVES = 2,200 BILLION BBLS

Figure A 11.2. Estimated world petroleum reserves.

Due to low success ratios encountered when exploring for commercial oil reservoirs, the industry soon found it necessary to develop a lower-cost technique for locating such reservoirs.

Thus, during the 1950s the offshore mobile unit underwent a rapid period of development. Mobile rigs made it possible to move on a location quickly, drill a well, and move to another location. With the advent of mobile drilling units, operations moved rapidly into deeper waters.

A full-scale exploratory well was drilled in 632 ft. of water off the coast of California several years ago. Right now there is a rig operating in approximately 640 ft. of water in the Santa Barbara Channel. Humble recently announced that the company would be drilling a well in that area in 1,300 ft. of water by year-end (see Figure A 11.3).

As indicated in Fig. A 11.4, the present producing depth record is

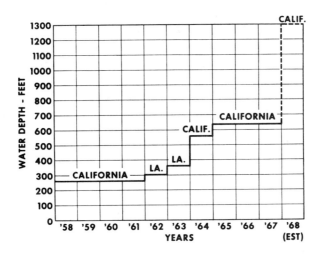

Figure A 11.3. Water depth records—exploratory wells.

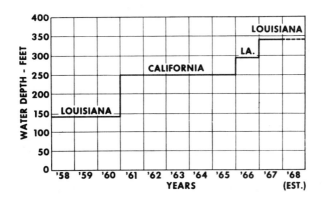

Figure A 11.4. Water depth records—producing wells.

340 ft. in a field off the coast of Louisiana. It is anticipated that new producing depth records will be established in the Santa Barbara Channel during the next few years.

RELATED STATISTICS

The statistics which follow provide an insight into the magnitude of past and current offshore operations:

One of the important factors in the expansion of offshore operations has been the development of marine seismic techniques for locating structures favorable for the occurrence of oil and gas. Although the number of marine crews operating offshore is only about 15 percent of the land crews, a marine crew can gather substantially more data than a land crew. (See Figures A 11.5 and A 11.6.)

One measure of the amount of data being obtained is the number of miles surveyed per year. As shown, the marine crews are currently surveying more miles than the land crews. The sudden spurt of marine activity in the 1962–65 period was caused by the rapid expansion of offshore leasing. During the past four years the total data gathered per year has remained essentially constant; however, the value of these data has been increased substantially through new data processing techniques. As you might expect, foreign marine seismic activity is currently greater than the total domestic activity—by a factor of 2 to 1. This trend is expected to continue. (See Figure A 11.7.)

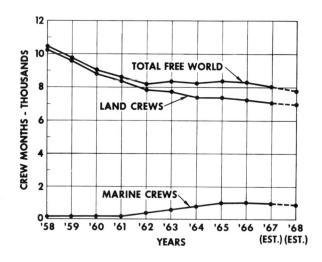

Figure A 11.5. Seismic surveys (land and marine)—crew months.

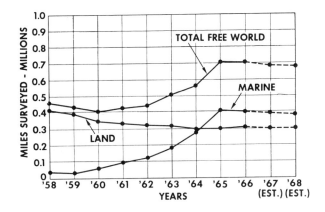

Figure A 11.6. Seismic surveys (land and marine)—miles surveyed.

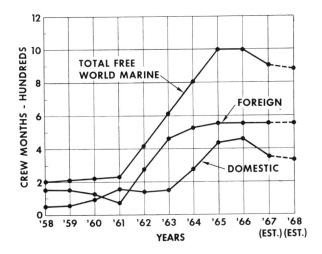

Figure A 11.7. Marine seismic surveys (domestic and foreign)—
crew months.

The number of offshore rigs operating throughout the world has more than doubled during the past five years. These mobile units are used primarily for exploration while the platform rigs are used to develop the reservoirs after they have been discovered. Some platforms use two rigs to reduce the time required to place the field on production. The number of rigs operating offshore in foreign areas has been increasing rapidly during the past few years. It is anticipated that these lines will cross in five years

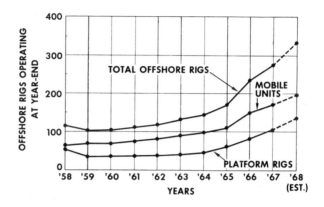

Figure A 11.8. Offshore drilling rigs (platform and mobile rigs).

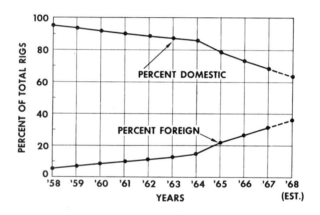

Figure A 11.9. Offshore drilling rigs (percent domestic and foreign).

or so as the overseas operations continue to expand. (See Figures A 11.8, A 11.9, and A 11.10.)

One important aspect of the move to deeper water has been the installation of pipe lines to connect the production platforms to shore. To date, the petroleum industry's experience in pipe laying has been limited to 340 ft. in medium-diameter pipe and 100 ft. in large-diameter pipe. In 1966, a French firm laid a 5-mile, 9⅝-in. line in the Mediterranean in water as deep as 1,080 ft. The so-called S-method used to lay this line was developed by Gaz de France in the early 1960s when they laid and recovered 9⅝-in. pipe in 8,500 ft. of water. Also, in the early 1960s, a jack-up construction barge was used to lay a 14-ft. diameter concrete line off California in 200 ft. of water.

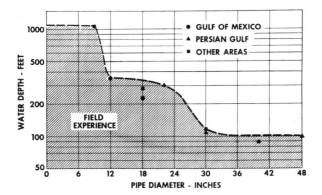

Figure A 11.10. Offshore pipelaying experience.

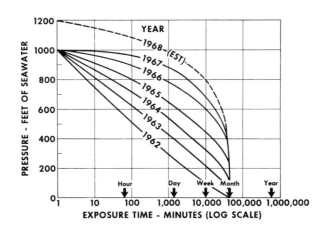

Figure A 11.11. Diving trends.

One very important trend in the over-all offshore picture has been the extension of diver capability into deep water. The trends are shown in Figure A 11.11. At the present time, the depth record for short-duration chamber dives stands at 1,100 ft. and at least one company has indicated that they will make a chamber dive to at least 1,200 ft. by year-end. It is interesting to note that Keller made a bounce dive to 1,000 ft. as early as 1962. Last year, saturated work dives were made to a depth of 636 ft. in the Gulf of Mexico. This series of dives was designed to determine the proficiency of divers while they were working on simulated oil field equipment, and the results were quite encouraging.

Saturation time is usually considered to be about 24 hours. By drawing a vertical line at this exposure time and cross-plotting, we come up with the following:

Saturation diving capabilities have been extended at a rate of approximately 125 ft. per year during the past six or seven years. This trend is expected to continue into the future; however, I am not in a position to predict what the ultimate saturation diving depth might be. (See Figure A 11.12.)

The number of offshore wells drilled per year in the United States has also increased rapidly during the past five years. There are about 1,600 offshore wells currently being drilled per year in U.S. waters. This includes wells being drilled from the Thums islands off Long Beach, Calif. (See Figure A 11.13.)

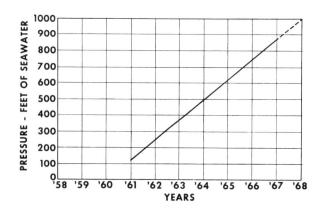

Figure A 11.12. Saturation diving trends.

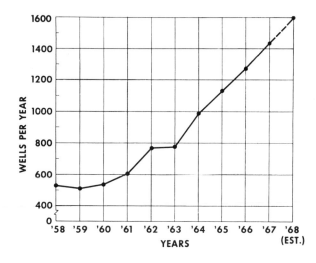

Figure A 11.13. Domestic offshore wells drilled per year.

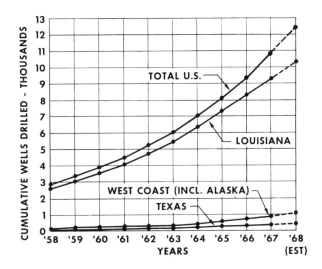

Figure A 11.14. Domestic offshore wells drilled to date.

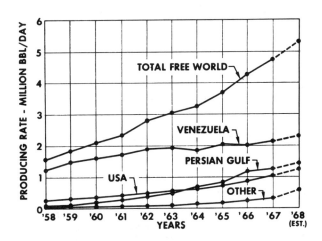

Figure A 11.15. Offshore oil producing rates.

Figure A 11.14 shows the cumulative number of domestic offshore wells. By year-end, the domestic total will be 12,500. Approximately 85 percent of these wells have been drilled in state and federal waters off the coast of Louisiana. Although accurate figures are not available, I estimate that the total number of foreign offshore wells drilled to date might stand at something like 7,500, giving a grand total of 20,000 wells for the Free World.

Figure A 11.15 shows the growth of Free World offshore oil production. The offshore growth rate has been approximately 15 percent per year compared with an over-all Free World growth rate for total petroleum production of 7–8 percent per year. Venezuela is the top offshore producing area; the Persian Gulf is second; and the United States is third (see Figure A 11.16).

The production of oil from offshore leases has increased from 10 up to 17 percent of total Free World production during the past 10 years. It is interesting to note that U.S. offshore production has increased from 4 up to 12 percent during the same period and is currently increasing at a higher rate than the rest of the Free World (see Figure A 11.17).

With this statistical background, I would now like to cover briefly some financial data on offshore operations.

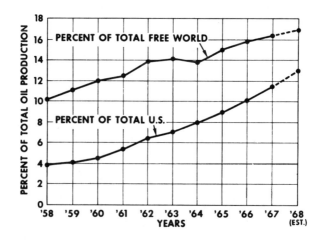

Figure A 11.16. Offshore oil production (percent of total production).

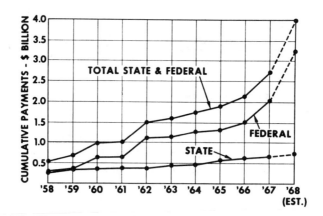

Figure A 11.17. Domestic offshore bonus and rental payments.

The total amount of money spent by the petroleum companies to obtain offshore leases in the United States has almost doubled in the last two years. This was primarily the result of three spectacular federal lease sales—a $500 million sale off Louisiana last year, plus two sales totaling approximately $600 million each this year—one in the Santa Barbara Channel and one in Texas waters. There was also a $500 million federal sale in 1962 and a $300 million federal sale in 1960 which added substantial amounts to these over-all totals. The current federal total stands at $3.3 billion and the state total is approximately $0.7 billion for a grand total of $4 billion. Cumulative royalties paid to date total $1.85 billion, about

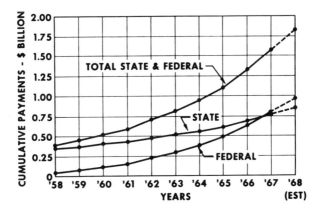

Figure A 11.18. Domestic offshore royalty payments.

BILLIONS OF DOLLARS

	1968 (EST)	CUMULATIVE (THRU 1968)
LEASE BONUS & RENTAL PAYMENTS	1.25	4.00
ROYALTY PAYMENTS	0.25	1.85
SEISMIC, GRAVITY & MAGNETIC SURVEYS	0.10	1.10
DRILLING & COMPLETING WELLS	0.35	3.10
PLATFORMS, PRODUCTION FACILITIES, & PIPELINES	0.25	1.85
OPERATING COSTS	0.15	0.85
TOTAL	2.25	12.75

Figure A 11.19. Domestic offshore expenditures.

equally divided between the states and federal government. (See Figure A 11.18.)

Here is one final bit of statistical information concerning domestic offshore expenditures. I estimate that in 1968 the petroleum industry will spend $2¼ billion in the domestic offshore arena. As I mentioned earlier, there have been two very large federal lease sales this year which account for about half of the total expenditures. Royalty payments to state and federal governments are running at a $250-million-per-year level. About $100 million is being spent in gathering, processing and analyzing geophysical data. A third of a billion dollars is being spent on drilling and completing both exploratory and development wells plus a quarter of a billion dollars on building platforms, production facilities and pipe lines to bring this production to market. Operating costs are running at an annual rate of $150 million. Total expenditures through the end of 1968 will reach an estimated $12.75 billion. I estimate that cumulative expenditures outside the United States might total $5 billion, giving a grand total of $18 billion for all Free World offshore development. (See Figure A 11.19.)

Appendix 12

Estimates of United States Domestic Demand for Liquid Hydrocarbons (1970–2000)*

1. Domestic production + imports = total U.S. supply.
2. Total U.S. supply, + or − inventory changes = total U.S. demand.
3. Total U.S. demand = domestic demand + export demand.
4. The concern is with the components of total U.S. supply.
5. *Figure A 12.1* shows published estimates of U.S. domestic demand. The values (points) appear in *Table A 12.1.*
6. U.S. domestic demand differs from total supply by the sum of exports and inventory change.
7. Exports and inventory change should approximate 50–70 million barrels a year during the period 1980–2000. This is about ½ to 1 percent of total supply.
8. The Department of the Interior forecast of July 1968 is almost the same as their 1965 forecast.
9. The high, medium and low RFF forecasts are the ones extending to year 2000 (high line goes off chart).†
10. Department of the Interior estimates for 1970–1980 ranges about one-fifth to one-third of the way between the medium and high RFF forecasts.
11. *Table A 12.4* shows the calculations of the new reserves requirements to meet the various RFF projections during 1981–2000, assuming:
    (a) imports = 20 percent of total supply, and
    (b) reserve/production ratio = 8 or 10:1 at end of century.

---

* From: *Petroleum Resources under the Ocean Floor, March 1969,* Washington; National Petroleum Council, 1969.

† Landsberg, Fischman, and Fisher, *Resources in America's Future* (1963), pp. 848–853. [RFF refers to Resources for the Future, Inc.]

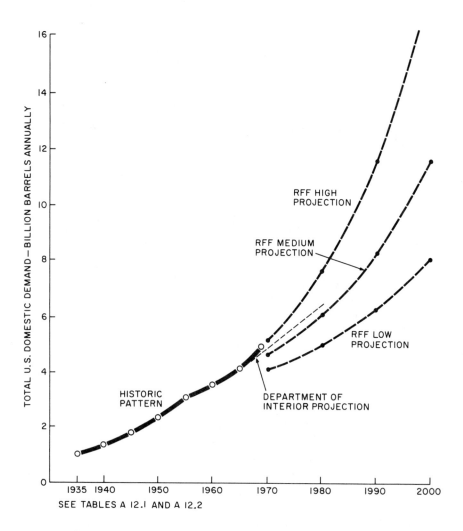

Figure A 12.1. Projection of United States total domestic demand
for liquid hydrocarbons.

TABLE A 12.1.—Forecasts of Domestic Demand Liquid Hydrocarbons (billion barrels per year)

	Resources in America's Future [a]			Department of the Interior	Bureau of Mines	Atlantic Richfield Company	Continental Oil Company [b]	Eastman Dillon Union Securities & Co.	Ebasco Services Incorporated	First National City Bank	Gulf Oil Corporation	Humble Oil & Refining Company	Mobil Oil Corporation	Standard Oil Company (Indiana)	Stanford Research Inst. SRI #226 — SRI #312	Texaco, Inc.	Texas Eastern Transmission Corporation	
	Low	Med.	High															
(Forecast date)				(1965)	(1966)	(1967)	(1965)	(1966)	(1965)	(1967)	(1966)	(1967)	(1965)	(1966)	(1965)	(1965)-(1967)	(1965)-(1961)	(1967)
Ref. Year	1960			1963	1965			1965	1964	1966	1966			1965		1966	1959	1965
Ref. Vol.	3.53			3.85	4.19			4.12	3.96	4.33	4.33			4.12		4.32	3.30	4.09
1970	4.07	4.62	5.25	4.89	—	—	—	—	5.03	4.80	4.94	—	—	—	—	—	4.63	4.85
1975	—	—	—	5.66	—	—	—	5.48	—	—	5.57	—	5.87	—	4.75	—	5.15	5.82
1980	5.03	6.09	7.66	6.42	6.66	6.20	6.39	—	6.22	—	6.24	—	—	6.20	5.48	5.66	5.72	6.93
1985	—	—	—	—	—	—	—	—	—	—	—	6.57	7.37	—	6.34	—	—	—
1990	6.31	8.21	11.62	—	—	—	—	—	—	—	—	—	—	—	—	—	—	8.21
1995	—	—	—	—	—	—	—	—	—	—	—	—	—	—	—	—	—	—
2000	8.10	11.59	18.32	—	—	—	—	—	—	—	—	—	—	—	—	—	—	—

a Landsberg et al., op. cit.
b Cram, Ira P.

TABLE A 12.2.—Department of the Interior, July 1968 Study[1]
[billions of barrels annually]

	Crude Oil Production	NGL Production	Total Liquids Production	Imports	Total Liquids Supply
1966	2.77	.59	3.36	.84	4.20
1967	2.87	.62	3.49	.87	4.36
1968	2.97	.65	3.62	.91	4.53
1969	3.07	.68	3.75	.94	4.69
1970	3.17	.71	3.88	.97	4.85
1971	3.27	.74	4.01	1.00	5.01
1972	3.37	.77	4.14	1.03	5.17
1973	3.47	.80	4.27	1.07	5.34
1974	3.57	.83	4.40	1.10	5.50
1975	3.67	.86	4.53	1.13	5.66
1976	3.76	.89	4.65	1.16	5.81
1977	3.86	.92	4.78	1.20	5.98
1978	3.96	.95	4.91	1.23	6.14
1979	4.06	.98	5.04	1.26	6.30
1980	4.16	1.01	5.18	1.20	6.46
Total (1966–80)	52.00	12.00	64.00	16.00	80.00

[1] See Figure A 12.1.

TABLE A 12.3—Calculations of New Reserves Requirements to Meet
Department of Interior Projections, 1965–80
[billions of barrels]

	Crude Oil	NGL	Total Liquids
Reserves at Jan. 1, 1966	38.9	8.0	46.9
Additions required, 15 years	56.0	11.5	67.5
Production, 15 years	52.0	12.0	64.0
Reserves at Jan. 1, 1981	42.9	7.5	50.4
Reserve/Production ratio 1965	14.4	13.7	14.2
Reserve/Production ratio 1980	10.2	7.4	9.7

TABLE A 12.4—Calculations of New Reserves Requirements to Meet RFF Projections, 1981–2000, Total Liquid Hydrocarbons
[billion barrels]

Case and Reserves at Jan. 1, 1981	Production 2000	Ratio of Reserves to Production at Jan. 1, 2001	Reserves at Jan. 1, 2001	Reserves 20 Years	Production 20 Years	Total New Reserve Requirement 20 Years
RFF—Low:						
50	6.5	10	65	15	103	118
50	6.5	8	52	2	103	105
RFF—Medium:						
50	9.3	10	93	43	136	179
50	9.3	8	74	24	136	160
RFF—High:						
50	14.6	10	146	96	197	293
50	14.6	8	117	67	197	264

Appendix 13

Public Laws 454 and 688

PUBLIC LAW 89–454

[89th Congress, S.944, June 17, 1966]

An Act to provide for a comprehensive, long-range, and coordinated national program in marine science, to establish a National Council on Marine Resources and Engineering Development, and a Commission on Marine Science, Engineering and Resources, and for other purposes

Be it enacted by the Senate and House of Representatives of the United States of America in Congress assembled, That this Act may be cited as the "Marine Resources and Engineering Development Act of 1966."

Declaration of Policy and Objectives

Sec. 2. (a) It is hereby declared to be the policy of the United States to develop, encourage, and maintain a coordinated, comprehensive, and long-range national program in marine science for the benefit of mankind to assist in protection of health and property, enhancement of commerce, transportation, and national security, rehabilitation of our commercial fisheries, and increased utilization of these and other resources.

(b) The marine science activities of the United States should be conducted so as to contribute to the following objectives:

1. The accelerated development of the resources of the marine environment.
2. The expansion of human knowledge of the marine environment.
3. The encouragement of private investment enterprise in exploration, technological development, marine commerce, and economic utilization of the resources of the marine environment.
4. The preservation of the role of the United States as a leader in marine science and resource development.
5. The advancement of education and training in marine science.
6. The development and improvement of the capabilities, performance, use, and efficiency of vehicles, equipment, and instruments

for use in exploration, research, surveys, the recovery of resources, and the transmission of energy in the marine environment.

7. The effective utilization of the scientific and engineering resources of the Nation, with close cooperation among all interested agencies, public and private, in order to avoid unnecessary duplication of effort, facilities, and equipment, or waste.

8. The cooperation by the United States with other nations and groups of nations and international organizations in marine science activities when such cooperation is in the national interest.

The National Council on Marine Resources and Engineering Development

Sec. 3. (a) There is hereby established, in the Executive Office of the President, the National Council on Marine Resources and Engineering Development (hereinafter called the "Council") which shall be composed of—

1. The Vice President, who shall be Chairman of the Council.
2. The Secretary of State.
3. The Secretary of the Navy.
4. The Secretary of the Interior.
5. The Secretary of Commerce.
6. The Chairman of the Atomic Energy Commission.
7. The Director of the National Science Foundation.
8. The Secretary of Health, Education, and Welfare.
9. The Secretary of the Treasury.

(b) The President may name to the Council such other officers and officials as he deems advisable.

(c) The President shall from time to time designate one of the members of the Council to preside over meetings of the Council during the absence, disability, or unavailability of the Chairman.

(d) Each member of the Council, except those designated pursuant to subsection (b), may designate any officer of his department or agency appointed with the advice and consent of the Senate to serve on the Council as his alternate in his unavoidable absence.

(e) The Council may employ a staff to be headed by a civilian executive secretary who shall be appointed by the President and shall receive compensation at a rate established by the President at not to exceed that of level II of the Federal Executive Salary Schedule. The executive secretary, subject to the direction of the Council, is authorized to appoint and fix the compensation of such personnel, including not more than seven persons who may be appointed without regard to civil service laws or the Classification Act of 1949 and compensated at not to exceed the highest rate of grade 18 of the General Schedule of the Classification Act of 1949, as amended, as may be necessary to perform such duties as may be prescribed by the President.

(f) The provisions of this Act with respect to the Council shall expire one hundred and twenty days after the submission of the final report of the Commission pursuant to section 5(h).

Responsibilities

Sec. 4. (a) In conformity with the provisions of section 2 of this Act, it shall be the duty of the President with the advice and assistance of the Council to—

1. Survey all significant marine science activities, including the policies, plans, programs, and accomplishments of all departments and agencies of the United States engaged in such activities;

2. Develop a comprehensive program of marine science activities, including, but not limited to, exploration, description and prediction of the marine environment, exploitation and conservation of the resources of the marine environment, marine engineering, studies of air-sea interaction, transmission of energy, and communications, to be conducted by departments and agencies of the United States, independently or in cooperation with such non-Federal organizations as States, institutions and industry;

3. Designate and fix responsibility for the conduct of the foregoing marine science activities by departments and agencies of the United States;

4. Insure cooperation and resolve differences arising among departments and agencies of the United States with respect to marine science activities under this Act, including differences as to whether a particular project is a marine science activity;

5. Undertake a comprehensive study, by contract or otherwise, of the legal problems arising out of the management, use, development, recovery, and control of the resources of the marine environment;

6. Establish long-range studies of the potential benefits to the United States economy, security, health, and welfare to be gained from marine resources, engineering, and science, and the costs involved in obtaining such benefits; and

7. Review annually all marine science activities conducted by departments and agencies of the United States in light of the policies, plans, programs, and priorities developed pursuant to this Act.

(b) In the planning and conduct of the coordinated Federal program the President and the Council shall utilize such staff, interagency, and non-Government advisory arrangements as they may find necessary and appropriate and shall consult with departments and agencies concerned with marine science activities and solicit the views of non-Federal organizations and individuals with capabilities in marine sciences.

Commission on Marine Science, Engineering, and Resources

Sec. 5. (a) The President shall establish a Commission on Marine Science, Engineering, and Resources (in this Act referred to as the "Commission"). The Commission shall be composed of fifteen members appointed by the President, including individuals drawn from Federal and State governments, industry, universities, laboratories and other institutions engaged in marine scientific or technological pursuits, but not more than five members shall be from the Federal Government. In addition the Commission shall have four advisory members appointed by the President from among the Members of the Senate and the House of Representatives. Such advisory members shall not participate, except in an advisory capacity, in the formulation of the findings and recommendations of the Commission. The President shall select a Chairman and Vice Chairman from among such fifteen members. The Vice Chairman shall act as Chairman in the latter's absence.

(b) The Commission shall make a comprehensive investigation and study of all aspects of marine science in order to recommend an overall plan for an adequate national oceanographic program that will meet the present and future national needs. The Commission shall undertake a review of existing and planned marine science activities of the United States in order to assess their adequacy in meeting the objectives set forth under section 2(b), including but not limited to the following:

1. Review the known and contemplated needs for natural resources from the marine environment to maintain our expanding national economy.
2. Review the surveys, applied research programs, and ocean engineering projects required to obtain the needed resources from the marine environment.
3. Review the existing national research programs to insure realistic and adequate support for basic oceanographic research that will enhance human welfare and scientific knowledge.
4. Review the existing oceanographic and ocean engineering programs, including education and technical training, to determine which programs are required to advance our national oceanographic competence and stature and which are not adequately supported.
5. Analyze the findings of the above reviews, including the economic

factors involved, and recommend an adequate national marine science program that will meet the present and future national needs without unnecessary duplication of effort.

6. Recommend a Governmental organizational plan with estimated cost.

(c) Members of the Commission appointed from outside the Government shall each receive $100 per diem when engaged in the actual performance of duties of the Commission and reimbursement of travel expenses, including per diem in lieu of subsistence, as authorized in section 5 of the Administrative Expenses Act of 1946, as amended (5 U.S.C. 73b–2), for persons employed intermittently. Members of the Commission appointed from within the Government shall serve without additional compensation to that received for their services to the Government but shall be reimbursed for travel expenses, including per diem in lieu of subsistence, as authorized in the Act of June 9, 1949, as amended (5 U.S.C. 835–842).

(d) The Commission shall appoint and fix the compensation of such personnel as it deems advisable in accordance with the civil service laws and the Classification Act of 1949, as amended. In addition, the Commission may secure temporary and intermittent services to the same extent as is authorized for the departments by section 15 of the Administrative Expenses Act of 1946 (60 Stat. 810) but at rates not to exceed $100 per diem for individuals.

(e) The Chairman of the Commission shall be responsible for (1) the assignment of duties and responsibilities among such personnel and their continuing supervision, and (2) the use and expenditures of funds available to the Commission. In carrying out the provisions of this subsection, the Chairman shall be governed by the general policies of the Commission with respect to the work to be accomplished by it and the timing thereof.

(f) Financial and administrative services (including those related to budgeting, accounting, financial reporting, personnel, and procurement) may be provided the Commission by the General Services Administration, for which payment shall be made in advance, or by reimbursement from funds of the Commission in such amounts as may be agreed upon by the Chairman of the Commission and the Administrator of General Services: *Provided,* That the regulations of the General Services Administration for the collection of indebtedness of personnel resulting from erroneous payments (5 U.S.C. 46d) shall apply to the collection of erroneous payments made to or on behalf of a Commission employee, and regulations of said Administrator for the administrative control of funds (31 U.S.C. 665 (g)) shall apply to appropriations of the Commission: *And provided further,* That the Commission shall not be required to prescribe such regulations.

(g) The Commission is authorized to secure directly from any executive department, agency, or independent instrumentality of the Government any information it deems necessary to carry out its functions under this Act; and each such department, agency, and instrumentality is authorized to cooperate with the Commission and, to the extent permitted by law, to furnish such information to the Commission, upon request made by the Chairman.

(h) The Commission shall submit to the President, via the Council, and to the Congress not later than eighteen months after the establishment of the Commission as provided in sub-section (a) of this section, a final report of its findings and recommendations. The Commission shall cease to exist thirty days after it has submitted its final report.

International Cooperation

Sec. 6. The Council, under the foreign policy guidance of the President and as he may request, shall coordinate a program of international cooperation in work done pursuant to this Act, pursuant to agreements made by the President with the advice and consent of the Senate.

Reports

Sec. 7. (a) The President shall transmit to the Congress in January of each year a report, which shall include (1) a comprehensive description of the activities and the accomplishments of all agencies and departments of the United States in the field of marine science during the preceding fiscal year, and (2) an evaluation of such activities and accomplishments in terms of the objectives set forth pursuant to this Act.

(b) Reports made under this section shall contain such recommendations for legislation as the President may consider necessary or desirable for the attainment of the objectives of this Act, and shall contain an estimate of funding requirements of each agency and department of the United States for marine science activities during the succeeding fiscal year.

Definitions

Sec. 8. For the purposes of this Act the term "marine science" shall be deemed to apply to oceanographic and scientific endeavors and disciplines, and engineering and technology in and with relation to the marine environment; and the term "marine environment" shall be deemed to include (a) the oceans, (b) the Continental Shelf of the United States, (c) the Great Lakes, (d) seabed and subsoil of the submarine areas adjacent to the coasts of the United States to the depth of two hundred meters, or beyond that limit, to where the depths of the superjacent waters admit of the exploitation of the natural resources of such areas, (e) the seabed and subsoil of similar submarine areas adjacent to the coasts of islands which comprise United States territory, and (f) the resources thereof.

Authorization

Sec. 9. There are hereby authorized to be appropriated such sums as may be necessary to carry out this Act, but sums appropriated for any one fiscal year shall not exceed $1,500,000.

Approved June 17, 1966.

Legislative History

House Reports: No. 1025 (Committee on Merchant Marine and Fisheries) and No. 1548 (committee of conference).
Senate Report No. 528 (Committee on Commerce).
Congressional Record:
  Volume 111 (1965):
    August 5, considered and passed Senate.
    September 20, considered and passed House, amended.
  Volume 112 (1966):
    May 26, House agreed to conference report.
    June 2, Senate agreed to conference report.

PUBLIC LAW 89–688

[89th Congress, H. R. 16559, October 15, 1966]

An Act to amend the Marine Resources and Engineering Development Act of 1966 to authorize the establishment and operation of sea grant colleges and programs by initiating and supporting programs of education and research in the various fields relating to the development of marine resources, and for other purposes

Be it enacted by the Senate and House of Representatives of the United States of America in Congress assembled, That the Marine Resources and Engineering Development Act of 1966 is amended by adding at the end thereof the following new title:

TITLE II—SEA GRANT COLLEGES AND PROGRAMS

Short Title

Sec. 201. This title may be cited as the "National Sea Grant College and Program Act of 1966."

Declaration of Purpose

Sec. 202. The Congress hereby finds and declares—

(a) that marine resources, including animal and vegetable life and mineral wealth, constitute a far-reaching and largely untapped asset of immense potential significance to the United States; and

(b) that it is in the national interest of the United States to develop the skilled manpower, including scientists, engineers, and technicians, and the facilities and equipment necessary for the exploitation of these resources; and

(c) that aquaculture, as with agriculture on land, and the gainful use of marine resources can substantially benefit the United States, and ultimately the people of the world, by providing greater economic opportunities, including expanded employment and commerce; the enjoyment and use of our marine resources; new sources of food; and new means for the development of marine resources; and

(d) that Federal support toward the establishment, development and operation of programs by sea grant colleges and Federal support of other sea grant programs designed to achieve the gainful use of marine resources, offer the best means of promoting programs toward the goals set forth in clauses (a), (b), and (c), and should be undertaken by the Federal Government; and

(e) that in view of the importance of achieving the earliest possible institution of significant national activities related to the development of marine resources, it is the purpose of this title to provide for the establishment of a program of sea grant colleges and education, training, and research in the fields of marine science, engineering, and related disciplines.

Grants and Contracts for Sea Grant Colleges and Programs

Sec. 203. (a) The provisions of this title shall be administered by the National Science Foundation (hereafter in this title referred to as the "Foundation").

(b) (1) For the purpose of carrying out this title, there is authorized to be appropriated to the Foundation for the fiscal year ending June 30, 1967, not to exceed the sum of $5,000,000, for the fiscal year ending June 30, 1968, not to exceed the sum of $15,000,000, and for each subsequent fiscal year only such sums as the Congress may hereafter specifically authorize by law.

(2) Amounts appropriated under this title are authorized to remain available until expended.

Marine Resources

Sec. 204. (a) In carrying out the provisions of this title the Foundation shall (1) consult with those experts engaged in pursuits in the various fields related to the development of marine resources and with all departments and agencies of the Federal Government (including the United States Office of Education in all matters relating to education) interested in, or affected by, activities in any such fields, and (2) seek advice and counsel from the National Council on Marine Resources and Engineering Development as provided by section 205 of this title.

(b) The Foundation shall exercise its authority under this title by—

1. initiating and supporting programs at sea grant colleges and other suitable institutes, laboratories, and public or private agencies for the education of participants in the various fields relating to the development of marine resources;

2. initiating and supporting necessary research programs in the various fields relating to the development of marine resources, with preference given to research aimed at practices, techniques, and design of equipment applicable to the development of marine resources; and

3. encouraging and developing programs consisting of instruction, practical demonstrations, publications, and otherwise, by sea grant colleges and other suitable institutes, laboratories, and public or private agencies through marine advisory programs with the object of imparting useful information to persons currently employed or interested in the various fields related to the development of marine resources, the scientific community, and the general public.

(c) Programs to carry out the purposes of this title shall be accomplished through contracts with, or grants to, suitable public or private institutions of higher education, institutes, laboratories, and public or private agencies which are engaged in, or concerned with, activities in the various fields related to the development of marine resources, for the establishment and operation by them of such programs.

(d) (1) The total amount of payments in any fiscal year under any grant to or contract with any participant in any program to be carried out by such participant under this title shall not exceed 66⅔ per centum of the total cost of such program. For purposes of computing the amount of the total cost of any such program furnished by any participant in any fiscal year, the Foundation shall include in such computation an amount equal to the reasonable value of any buildings, facilities, equipment, supplies, or services provided by such participant with respect to such program (but not the cost or value of land or of Federal contributions).

(2) No portion of any payment by the Foundation to any participant in any program to be carried out under this title shall be applied to the purchase or rental of any land or the rental, purchase, construction, preservation, or repair of any building, dock, or vessel.

(3) The total amount of payments in any fiscal year by the Foundation to participants within any State shall not exceed 15 per centum of the total amount appropriated to the Foundation for the purposes of this title for such fiscal year.

(e) In allocating funds appropriated in any fiscal year for the purposes of this title the Foundation shall endeavor to achieve maximum participation by sea grant colleges and other suitable institutes, laboratories, and public or private agencies throughout the United States, consistent with the purposes of this title.

(f) In carrying out its functions under this title, the Foundation shall attempt to support programs in such a manner as to supplement and not duplicate or overlap any existing and related Government activities.

(g) Except as otherwise provided in this title, the Foundation shall, in carrying out its functions under this title, have the same powers and authority it has under the National Science Foundation Act of 1950 to carry out its functions under that Act.

(h) The head of each department, agency, or instrumentality of the Federal Government is authorized, upon request of the Foundation, to make available to the Foundation, from time to time, on a reimbursable basis, such personnel, services, and facilities as may be necessary to assist the Foundation in carrying out its functions under this title.

(i) For the purposes of this title—

1. The term "development of marine resources" means scientific endeavors relating to the marine environment, including, but not limited to, the fields oriented toward the development, conservation, or economic utilization of the physical, chemical, geological, and biological resources of the marine environment; the fields of marine commerce and marine engineering; the fields relating to exploration or research in, the recovery of natural resources from, and the transmission of energy in, the marine environment; the fields of oceanography and oceanology; and the fields with respect to the study of the economic, legal, medical, or sociological problems arising out of the management, use, development, recovery, and control of the natural resources of the marine environment;

2. The term "marine environment" means the oceans; the Continental Shelf of the United States; the Great Lakes; the seabed and subsoil of the submarine areas adjacent to the coasts of the United States to the depth of two hundred meters, or beyond that limit,

to where the depths of the superjacent waters admit of the exploitation of the natural resources of the area; the seabed and subsoil of similar submarine areas adjacent to the coasts of islands which comprise United States territory; and the natural resources thereof;

3. The term "sea grant college" means any suitable public or private institution of higher education supported pursuant to the purposes of this title which has major programs devoted to increasing our Nation's utilization of the world's marine resources; and

4. The term "sea grant program" means (A) any activities of education or research related to the development of marine resources supported by the Foundation by contracts with or grants to institutions of higher education either initiating, or developing existing, programs in fields related to the purposes of this title, (B) any activities of education or research related to the development of marine resources supported by the Foundation by contracts with or grants to suitable institutes, laboratories, and public or private agencies, and (C) any programs of advisory services oriented toward imparting information in fields related to the development of marine resources supported by the Foundation by contracts with or grants to suitable institutes, laboratories, and public or private agencies.

Advisory Functions

Sec. 205. The National Council on Marine Resources and Engineering Development established by section 3 of title I of this Act shall, as the President may request—

1. Advise the Foundation with respect to the policies, procedures, and operations of the Foundation in carrying out its functions under this title;

2. Provide policy guidance to the Foundation with respect to contracts or grants in support of programs conducted pursuant to this title, and make such recommendations thereon to the Foundation as may be appropriate; and

3. Submit an annual report on its activities and its recommendations under this section to the Speaker of the House of Representatives, the Committee on Merchant Marine and Fisheries of the House of Representatives, the President of the Senate, and the Committee on Labor and Public Welfare of the Senate.

Sec. 2. (a) The Marine Resources and Engineering Development Act of 1966 is amended by striking out the first section and inserting in lieu thereof the following:

TITLE I—MARINE RESOURCES AND ENGINEERING DEVELOPMENT

Short Title

Section 1. This title may be cited as the "Marine Resources and Engineering Development Act of 1966."

(b) Such Act is further amended by striking out "this Act" the first place it appears in section 4(a), and also each place it appears in sections 5(a), 8, and 9, and inserting in lieu thereof in each such place "this title."

Approved October 15, 1966.

Legislative History

House reports: No. 1795 (Commission on Merchant Marine and Fisheries) and No. 2156 (commission of conference).

Senate Report No. 1307 accompanying S. 2439 (Commission on Labor and Public Welfare).

Congressional Record, volume 112 (1966):

September 13: Considered and passed House.

September 14: Considered and passed Senate, amended, in lieu of S. 2439.

September 30: Senate agreed to conference report.

October 4: House agreed to conference report.

Appendix 14

Draft Resolutions Submitted and Principles Supported by the United States in the Ad Hoc Seabed Committee, June 28, 1968

A. U.S. DRAFT RESOLUTION CONTAINING STATEMENT OF PRINCIPLES CONCERNING THE DEEP OCEAN FLOOR

The General Assembly,

Desiring to encourage the exploration, use and development of the deep ocean floor to the fullest extent possible for the benefit and in the interest of all mankind,

Believing that such exploration and use of the deep ocean floor will contribute to international co-operation and understanding,

Convinced that no nation, regardless of geographical location, level of economic development, or technological capability, should be denied the opportunity to participate in the exploration and use of the deep ocean floor,

Conscious of the importance of promoting the general welfare of all peoples, and of furthering scientific study and the conservation of resources,

Reaffirming the traditional freedoms of the high seas under international law,

Recalling its resolution 2340 (XXII) of 18 December 1967,

Commends to States for their guidance the following principles concerning the deep ocean floor:

1. No State may claim or exercise sovereignty or sovereign rights over any part of the deep ocean floor. There shall be no discrimination in the availability of the deep ocean floor for exploration and use by all States and their nationals in accordance with international law;

2. There shall be established, as soon as practicable, internationally agreed arrangements governing the exploitation of resources of the deep ocean floor. These arrangements shall reflect the other principles contained in this Statement of Principles concerning the Deep Ocean Floor and shall include provision for:

(a) The orderly development of resources of the deep ocean floor in a manner reflecting the interest of the international community in the development of these resources;

(b) Conditions conducive to the making of investments necessary for the exploration and exploitation of resources of the deep ocean floor;

(c) Dedication as feasible and practicable of a portion of the value of the resources recovered from the deep ocean floor to international community purposes; and

(d) Accommodation among the commercial and other uses of the deep ocean floor and marine environment;

3. Taking into account the Geneva Convention of 1958 on the Continental Shelf, there shall be established, as soon as practicable, an internationally agreed precise boundary for the deep ocean floor—the seabed and subsoil beyond that over which coastal States may exercise sovereign rights for the purpose of exploration and exploitation of its natural resources; exploitation of the natural resources of the ocean floor that occurs prior to establishment of the boundary shall be understood not to prejudice its location, regardless of whether the coastal State considers the exploitation to have occurred on its "continental shelf";

4. States and their nationals shall conduct their activities on the deep ocean floor in accordance with international law, including the Charter of the United Nations, and in the interest of maintaining international peace and security and promoting international co-operation, scientific knowledge, and economic development;

5. In order to further international co-operation in the scientific investigation of the deep ocean floor, States shall:

(a) Disseminate, in a timely fashion, plans for and results of national scientific programmes concerning the deep ocean floor;

(b) Encourage their nationals to follow similar practices concerning dissemination of such information;

(c) Encourage co-operative scientific activities regarding the deep ocean floor by personnel of different States;

6. In the exploration and use of the deep ocean floor States and their nationals:

(a) Shall have reasonable regard for the interests of other States and their nationals;

(b) Shall avoid unjustifiable interference with the exercise of the freedom of the high seas by other States and their nationals, or with the conservation of the living resources of the seas, and any interference with fundamental scientific research carried out with the intention of open publication;

(c) Shall adopt appropriate safeguards so as to minimize pollution of the seas and disturbance of the existing biological, chemical and physical processes and balances; each State shall provide timely announcement and any necessary amplifying information of any marine activity or experiment planned by it or its nationals that could harmfully interfere with the activities of any other State or its nationals in the exploration and use of the deep ocean floor. A State which has reason to believe that a marine activity or experiment planned by another State or its nationals could harmfully interfere with its activities or those of its nationals in the exploration and use of the deep ocean floor may request consultation concerning the activity or experiment;

7. States and their nationals shall render all possible assistance to one another in the event of accident, distress or emergency arising out of activities on the deep ocean floor.

B. U.S. DRAFT RESOLUTION ON PREVENTING THE EMPLACEMENT OF WEAPONS OF MASS DESTRUCTION ON THE SEABED AND OCEAN FLOOR

The General Assembly,

Desiring that workable arms limitation measures be achieved that will enhance the peace and security of all nations and bring the world nearer to general and complete disarmament,

Requests the Eighteen-Nation Committee on Disarmament to take up the question of arms limitation on the seabed and ocean floor with a view to defining those factors vital to a workable, verifiable and effective international agreement which would prevent the use of this new environment for the emplacement of weapons of mass destruction.

C. U.S. DRAFT RESOLUTION ON THE INTERNATIONAL DECADE OF OCEAN EXPLORATION

The General Assembly,

Recalling its concern for ascertaining practical means to promote international co-operation in the exploration, conservation and use of the seabed and the ocean floor, and the subsoil thereof, as manifested in its resolution 2340 (XXII),

Recalling as well that in its resolution 2172 (XXI) it requested that the Secretary-General prepare proposals for ensuring the most effective arrangements for an expanded programme of international co-operation to assist in a better understanding of the marine environment through science, and for initiating and strengthening marine education and training programmes,

Recalling further the proposals made by the Secretary-General in his report (E/4487) pursuant to resolution 2172 (XXI),

Noting that the Bureau and Consultative Council of the Intergovernmental Oceanographic Commission of UNESCO considered the proposed International Decade of Ocean Exploration a useful initiative for broadening and accelerating investigations of the oceans and for strengthening international co-operation,

Noting also the recommendation adopted by the Economic and Social Council on 2 August 1968, inviting the General Assembly to endorse the concept of a co-ordinated long-term programme of oceanographic research, taking into account such initiatives as the proposal for an International Decade of Ocean Exploration and international programmes already considered, approved and adopted by the Intergovernmental Oceanographic Commission for implementation in co-operation with other specialized agencies,

Aware of the consideration given to the proposal in the *Ad Hoc* Committee to Study the Peaceful Uses of the Seabed and the Ocean Floor, arising from the important contribution which the Decade would make to scientific research and exploration of the seabed and deep ocean floor.

1. *Welcomes* and commends to Member States the concept of an International Decade of Ocean Exploration to be undertaken within the framework of a long-term programme of research and exploration under the general aegis of the United Nations;

2. *Invites* interested Member States to formulate proposals for national and international scientific programmes and agreed activities to be undertaken during the Decade with due regard to the interests of developing countries, to transmit these proposals to the Intergovernmental Oceanographic Commission, and to begin such activities as soon as practicable;

3. *Urges* Member States to publish as soon as practicable the results of activities which they will have undertaken within the framework of the Decade and at the same time to communicate these results to the Intergovernmental Oceanographic Commission;

4. *Requests* the Intergovernmental Oceanographic Commission:

   (a) To further and co-ordinate, in co-operation with other interested agencies, and expanded, accelerated, long-term and sustained programme of world-wide exploration of the oceans and their resources of which the Decade will be an element, including international agency programmes, expanded international exchange of data from national programmes, and international efforts to strengthen the research capabilities of all interested nations;

(b) To report through appropriate channels to the twenty-fourth session of the General Assembly on the progress made in ocean activities undertaken pursuant to this resolution.

D. DRAFT STATEMENT OF AGREED PRINCIPLES PROPOSED FOR SUBMISSION TO THE GENERAL ASSEMBLY AND SUPPORTED BY THE UNITED STATES

(1) There is an area of the seabed and ocean floor and the subsoil thereof, underlying the high seas, which lies beyond the limits of national jurisdiction (hereinafter described as "this area");

(2) Taking into account relevant dispositions of international law, there should be agreed a precise boundary for this area;

(3) There should be agreed, as soon as practicable, an international régime governing the exploitation of resources of this area;

(4) No State may claim or exercise sovereign rights over any part of this area, and no part of it is subject to national appropriation by claim of sovereignty, by use or occupation, or by any other means;

(5) Exploration and use of this area shall be carried on for the benefit and in the interests of all mankind, taking into account the special needs of the developing countries;

(6) This area shall be reserved exclusively for peaceful purposes;

(7) Activities in this area shall be conducted in accordance with international law, including the Charter of the United Nations. Activities in this area shall not infringe upon the freedoms of the high seas.

Appendix 15

Treaty on the Prohibition of the Emplacement of Nuclear Weapons and Other Weapons of Mass Destruction on the Seabed and the Ocean Floor and in the Subsoil Thereof

The States Parties to this Treaty,

Recognizing the common interest of mankind in the progress of the exploration and use of the seabed and the ocean floor for peaceful purposes,

Considering that the prevention of a nuclear arms race on the seabed and the ocean floor serves the interests of maintaining world peace, reduces international tensions and strengthens friendly relations among States,

Convinced that this Treaty constitutes a step towards a treaty on general and complete disarmament under strict and effective international control, and determined to continue negotiations to this end,

Convinced that this Treaty will further the purposes and principles of the Charter of the United Nations, in a manner consistent with the principles of international law and without infringing the freedoms of the high seas;

Have agreed as follows:

ARTICLE I

1. The States Parties to this Treaty undertake not to emplant or emplace on the seabed and the ocean floor and in the subsoil thereof beyond the outer limit of a seabed zone, as defined in article II, any nuclear weapons or any other types of weapons of mass destruction as well as structures, launching installations or any other facilities specifically designed for storing, testing or using such weapons.

2. The undertakings of paragraph 1 of this article shall also apply to the seabed zone referred to in the same paragraph, except that within such seabed zone, they shall not apply either to the coastal State or to the seabed beneath its territorial waters.

3. The States Parties to this Treaty undertake not to assist, encourage or induce any State to carry out activities referred to in paragraph 1 of this article and not to participate in any other way in such actions.

ARTICLE II

For the purpose of this Treaty, the outer limit of the seabed zone referred to in article I shall be coterminous with the twelve-mile outer limit of the zone referred to in part II of the Convention on the Territorial Sea and the Contiguous Zone, signed at Geneva on April 29, 1958, and shall be measured in accordance with the provisions of part I, section II, of that Convention and in accordance with international law.

ARTICLE III

1. In order to promote the objectives of and insure compliance with the provisions of this Treaty, each State Party to the Treaty shall have the right to verify through observation the activities of other States Parties to the Treaty on the seabed and the ocean floor and in the subsoil thereof beyond the zone referred to in article I, provided that observation does not interfere with such activities.

2. If after such observation reasonable doubts remain concerning the fulfillment of the obligations assumed under the Treaty, the State Party having such doubts and the State Party that is responsible for the activities giving rise to the doubts shall consult with a view to removing the doubts. If the doubts persist, the State Party having such doubts shall notify the other States Parties, and the Parties concerned shall cooperate on such further procedures for verification as may be agreed, including appropriate inspection of objects, structures, installations or other facilities that reasonably may be expected to be of a kind described in article I. The Parties in the region of the activities, including any coastal State, and any other Party so requesting, shall be entitled to participate in such consultation and cooperation. After completion of the further procedures for verification, an appropriate report shall be circulated to other Parties by the Party that initiated such procedures.

3. If the State responsible for the activities giving rise to the reasonable doubts is not identifiable by observation to the object, structure, installation or other facility, the State Party having such doubts shall notify and make appropriate inquiries of States Parties in the region of the activities and of any other State Party. If it is ascertained through these inquiries that a particular State Party is responsible for the activities, that State Party

shall consult and cooperate with other Parties as provided in paragraph 2 of this article. If the identity of the State responsible for the activities cannot be ascertained through these inquiries, then further verification procedures, including inspection, may be undertaken by the inquiring State Party, which shall invite the participation of the Parties in the region of the activities, including any coastal State, and of any other Party desiring to cooperate.

4. If consultation and cooperation pursuant to paragraph 2 and 3 of this article have not removed the doubts concerning the activities and there remains a serious question concerning fulfillment of the obligations assumed under this Treaty, a State Party may, in accordance with the provisions of the Charter of the United Nations, refer the matter to the Security Council, which may take action in accordance with the Charter.

5. Verification pursuant to this article may be undertaken by any State Party using its own means, or with the full or partial assistance of any other State Party, or through appropriate international procedures within the framework of the United Nations and in accordance with its Charter.

6. Verification activities pursuant to this Treaty shall not interfere with activities of other States Parties and shall be conducted with due regard for rights recognized under international law, including the freedoms of the high seas and the rights of coastal States with respect to the exploration and exploitation of their continental shelves.

ARTICLE IV

Nothing in this Treaty shall be interpreted as supporting or prejudicing the position of any State Party with respect to existing international conventions, including the 1958 Convention on the Territorial Sea and the Contiguous Zone, or with respect to rights or claims which such State Party may assert, or with respect to recognition or nonrecognition of rights or claims asserted by any other State, related to waters off its coasts, including, *inter alia,* territorial seas and contiguous zones, or to the seabed and the ocean floor, including continental shelves.

ARTICLE V

The Parties to this Treaty undertake to continue negotiations in good faith concerning further measures in the field of disarmament for the prevention of an arms race on the seabed, the ocean floor and the subsoil thereof.

ARTICLE VI

Any State Party may propose amendments to this Treaty. Amendments shall enter into force for each State Party accepting the amendments upon their acceptance by a majority of the States Parties to the Treaty and, thereafter, for each remaining State Party on the date of acceptance by it.

ARTICLE VII

Five years after the entry into force of this Treaty, a conference of Parties to the Treaty shall be held at Geneva, Switzerland, in order to review the operation of this Treaty with a view to assuring that the purposes of the preamble and the provisions of the Treaty are being realized. Such review shall take into account any relevant technological developments. The review conference shall determine, in accordance with the views of a majority of those Parties attending, whether and when an additional review conference shall be convened.

ARTICLE VIII

Each State Party to this Treaty shall in exercising its national sovereignty have the right to withdraw from this Treaty if it decides that extraordinary events related to the subject matter of this Treaty have jeopardized the supreme interests of its country. It shall give notice of such withdrawal to all other States Parties to the Treaty and to the United Nations Security Council three months in advance. Such notice shall include a statement of the extraordinary events it considers to have jeopardized its supreme interests.

ARTICLE IX

The provisions of this Treaty shall in no way affect the obligations assumed by States Parties to the Treaty under international instruments establishing zones free from nuclear weapons.

ARTICLE X

1. This Treaty shall be open for signature to all States. Any State which does not sign the Treaty before its entry into force in accordance with paragraph 3 of this article may accede to it at any time.

2. This Treaty shall be subject to ratification by signatory States. Instruments of ratification and of accession shall be deposited with the Governments of the United States of America, the United Kingdom of Great Britain and Northern Ireland, and the Union of Soviet Socialist Republics, which are hereby designated the Depositary Governments.

3. This Treaty shall enter into force after the deposit of instruments of ratification by twenty-two Governments, including the Governments designated as Depositary Governments of this Treaty.

4. For States whose instruments of ratification or accession are deposited after the entry into force of this Treaty, it shall enter into force on the date of the deposit of their instruments of ratification or accession.

5. The Depositary Governments shall promptly inform the Governments of all signatory and acceding States of the date of each signature, of the date of deposit of each instrument of ratification or of accession, of the date of the entry into force of this Treaty, and of the receipt of other notices.

6. This Treaty shall be registered by the Depositary Governments pursuant to Article 102 of the Charter of the United Nations.

ARTICLE XI

This Treaty, the English, Russian, French, Spanish and Chinese texts of which are equally authentic, shall be deposited in the archives of the Depositary Governments. Duly certified copies of this Treaty shall be transmitted by the Depositary Governments to the Governments of the States signatory and acceding thereto.

IN WITNESS WHEREOF the undersigned, being duly authorized thereto, have signed this Treaty.

DONE in triplicate, at the cities of Washington, London and Moscow, this eleventh day of February, one thousand nine hundred seventy-one.

For the United States of America:
William P. Rogers
James F. Leonard
For the United Kingdom of Great Britain and Northern Ireland:
Cromer
For the Union of Soviet Socialist Republics:
A. Dobrynin
For Finland:
Olavi Munkki

For Costa Rica:
R. A. Zuñiga
For Liberia:
S. Edward Peal
For New Zealand:
R. L. Jermyn
For Nepal:
Kul Shekhar Sharma
For the Khmer Republic:
Sonn
For Luxembourg:
Jean Wagner

For Jordan:
A. Sharaf
For Sweden:
Hubert De Besche
For Cyprus:
Zenon Rossides
For Bulgaria:
D. L. Guerassimov
For Burma:
San Maung
For Hungary:
Nagy János
For Iceland:
Hördur Helgason
For Canada:
Marcel Cadieux
For Ethiopia:
Minasse Haile
For the Republic of Korea:
Dong Jo Kim
[Romanization]
For Laos:
Lane Pathammavong
For Guatemala:
J. Asensio Wunderlich
For Denmark:
Torben Rønne
For Honduras:
Roberto Galvez
For the Republic of China:
Chow Shukai
For Ireland:
W. Warnock
For the Niger:
A. Joseph [Joseph Amina]
For Romania:
Corneliu Bogdan
*For The Central African
Republic:*
R. Guerillot
For The Dominican Republic:
S. Ortiz

For Mali:
S. Traoré
For Czechoslovakia:
D. Rohal-Ilkiv
For Tanzania:
Shilam
[G. M. Rutabanzibwa]
For Poland:
Jerzy Michalowski
For Japan:
Nobuhiko Ushiba
For Afghanistan:
A. Malikyar
For Morocco:
Abdeslam Tadlaoui
For Nicaragua:
Guillermo Sevilla-Sacasa
For Malta:
Arvid Pardo
For Lebanon:
N. Kabbani
For Republic of Viet-Nam:
Bui-Diem
For Iran:
Dr. A. Aslan Afshar
For Austria:
Gruber
For Ghana:
E. Moses Debrah
For Tunisia:
S. El Goulli
For Norway:
Arne Gunneng
For Australia:
John Ryan
For The Netherlands:
R. B. Van Lynden
For Mauritius:
Pierre Guy Girald
Balancy
For South Africa:
H. L. T. Taswell

For Switzerland:
Felix Schnyder

For Belgium:
Walter Loridan

For Burundi:
Nsanze Térence

For Swaziland:
S. T. M. Sukati

For Rwanda:
Fidèle Nkundabagenzi

For Italy:
Egidio Ortona

For Uruguay:
H. Luisi

For Bolivia:
A. S. de Lozada

For Botswana:
Linchwe II

For Colombia:
D. Botero B

For Guinea:
F. Keita

For Panama:
J. A. de la Ossa

For Greece:
B. Vitsaxis
12 of February 1971.

For Paraguay:
Roque J. Avila
23 February 1971.

For Sierra Leone:
John J. Akar
24th February 1971.

For Turkey:
Melih Esenbel
Feb. 25, 1971.

For Yugoslavia:
B. Crnobrnja
2 of March 1971.

For Senegal:
Cheikh Fall
17th of March 1971.

For Dahomey:
Wilfrid de Souza
le 18 Mars 1971.

Appendix 16

United States Oceans Policy*

STATEMENT BY THE PRESIDENT. MAY 23, 1970

The nations of the world are now facing decisions of momentous importance to man's use of the oceans for decades ahead. At issue is whether the oceans will be used rationally and equitably and for the benefit of mankind or whether they will become an arena of unrestrained exploitation and conflicting jurisdictional claims in which even the most advantaged states will be losers.

The issue arises now—and with urgency—because nations have grown increasingly conscious of the wealth to be exploited from the seabeds and throughout the waters above, and because they are also becoming apprehensive of the ecological hazards of unregulated use of the oceans and seabeds. The stark fact is that the law of the sea is inadequate to meet the needs of modern technology and the concerns of the international community. If it is not modernized multilaterally, unilateral action and international conflict are inevitable.

This is the time, then for all nations to set about resolving the basic issues of the future regime for the oceans—and to resolve it in a way that redounds to the general benefit in the era of intensive exploitation that lies ahead. The United States as a maritime power and a leader in ocean technology to unlock the riches of the ocean has a special responsibility to move this effort forward.

Therefore, I am today proposing that all nations adopt as soon as possible a treaty under which they would renounce all national claims over the natural resources of the seabed beyond the point where the high seas reach a depth of 200 meters (218.8 yards) and would agree to regard these resources as the common heritage of mankind.

The treaty should establish an international regime for the exploitation of seabed resources beyond this limit. The regime should include the collection of substantial mineral royalties to be used for international community purposes, particularly economic assistance to developing countries.

* Source: *Weekly Compilation of Presidential Documents,* Vol. 6, No. 21, May 25, 1970, pages 677–678.

The regime should also establish general rules to prevent unreasonable interference with other uses of the ocean, to protect the ocean from pollution, to assure the integrity of the investment necessary for such exploitation, and to provide for peaceful and compulsory settlement of disputes.

I propose two types of machinery for authorizing exploitation of sea-bed resources beyond a depth of 200 meters.

First, I propose that coastal nations act as trustees for the international community in an international trusteeship zone comprised of the continental margins beyond a depth of 200 meters off their coasts. In return, each coastal state would receive a share of the international revenues from the zone in which it acts as trustee and could impose additional taxes if it deemed this desirable.

As a second step, agreed international machinery would authorize and regulate exploration and use of seabed resources beyond the continental margins.

The United States will introduce specific proposals at the next meeting of the United Nations Seabeds Committee to carry out these objectives.

Although I hope agreement on such steps can be reached quickly, the negotiation of such a complex treaty may take some time. I do not, however, believe it is either necessary or desirable to try to halt exploration and exploitation of the seabeds beyond a depth of 200 meters during the negotiating process.

Accordingly, I call on other nations to join the United States in an interim policy. I suggest that all permits for exploration and exploitation of the seabeds beyond 200 meters be issued subject to the international regime to be agreed upon. The regime should accordingly include due protection for the integrity of investments made in the interim period. A substantial portion of the revenues derived by a state from exploitation beyond 200 meters during this interim period should be turned over to an appropriate international development agency for assistance to developing countries. I plan to seek appropriate congressional action to make such funds available as soon as a sufficient number of other states also indicate their willingness to do so.

I will propose necessary changes in the domestic import and tax laws and regulations of the United States to assure that our own laws and regulations do not discriminate against U.S. nationals operating in the trusteeship zone off our coast or under the authority of the international machinery to be established.

It is equally important to assure unfettered and harmonious use of the oceans as an avenue of commerce and transportation, and as a source of food. For this reason the United States is currently engaged with other states in an effort to obtain a new law of the sea treaty. This treaty would establish a 12-mile limit for territorial seas and provide for free transit through international straits. It would also accommodate the problems of developing countries and other nations regarding the conservation and use of the living resources of the high seas.

I believe that these proposals are essential to the interests of all nations, rich and poor, coastal and landlocked, regardless of their political systems. If they result in international agreements, we can save over two-thirds of the earth's surface from national conflict and rivalry, protect it from pollution, and put it to use for the benefit of all. This would be a fitting achievement for this 25th anniversary year of the United Nations.

Appendix 17

Draft United Nations Convention on the International Seabed Area, August 3, 1970*

Working Paper

The attached draft of a United Nations Convention on the International Seabed Area is submitted by the United States Government as a working paper for discussion purposes.

The draft Convention and its Appendices raise a number of questions with respect to which further detailed study is clearly necessary and do not necessarily represent the definitive views of the United States Government. The Appendices in particular are included solely by way of example.

UNITED NATIONS CONVENTION ON THE INTERNATIONAL SEABED AREA

Chapter I—Basic Principles

ARTICLE 1

1. The International Seabed Area shall be the common heritage of all mankind.

2. The International Seabed Area shall comprise all areas of the seabed and subsoil of the high seas† seaward of the 200 meter isobath adjacent to the coast of continents and islands.

3. Each Contracting Party shall permanently delineate the precise boundary of the International Seabed Area off its coast by straight lines not exceeding 60 nautical miles in length, following the general direction of the limit specified in paragraph 2. Such lines shall connect fixed points at the

---

* U.N. Doc. A/AC. 138/25.

† The United States has simultaneously proposed an international Convention (Appendix 18) which would *inter alia,* fix the boundary between the territorial sea and the high seas at a maximum distance of 12 nautical miles from the coast.

limit specified in paragraph 2, defined permanently by coordinates of latitude and longitude. Areas between or landward of such points may be deeper than 200 meters. Where a trench or trough deeper than 200 meters transects an area less than 200 meters in depth, a straight boundary line more than 60 nautical miles in length, but not exceeding the lesser of one fourth of the length of that part of the trench or trough transecting the area 200 meters in depth or 120 nautical miles, may be drawn across the trench or trough.

4. Each Contracting Party shall submit the description of the boundary to the International Seabed Boundary Review Commission within five years of the entry into force of this Convention for such Contracting Party. Boundaries not accepted by the Commission and not resolved by negotiation between the Commission and the Contracting Party within one year shall be submitted by the Commission to the Tribunal in accordance with Section E of Chapter IV.

5. Nothing in this Article shall affect any agreement or prejudice the position of any Contracting Party with respect to the delimitation of boundaries between opposite or adjacent States in seabed landward of the International Seabed Area, or with respect to any delimitation pursuant to Article 30.

ARTICLE 2

1. No State may claim or exercise sovereign rights over any part of the International Seabed Area or its resources. Each Contracting Party agrees not to recognize any such claim or exercise of sovereignty or sovereign rights.

2. No State has, nor may it acquire, any right, title, or interest in the International Seabed Area or its resources except as provided in this Convention.*

ARTICLE 3

The International Seabed Area shall be open to use by all States, without discrimination, except as otherwise provided in this Convention.

ARTICLE 4

The International Seabed Area shall be reserved exclusively for peaceful purposes.

ARTICLE 5

1. The International Seabed Resource Authority shall use revenues it derives from the exploration and exploitation of the mineral resources of

---

* NOTE.—The preceding Article is not intended to imply that States do not currently have rights under, or consistent with, the 1958 Geneva Convention on the Continental Shelf.

the International Seabed Area for the benefit of all mankind, particularly to promote the economic advancement of developing States Parties to this Convention, irrespective of their geographic location. Payments to the Authority shall be established at levels designed to ensure that they make a continuing and substantial contribution to such economic advancement, bearing in mind the need to encourage investment in exploration and exploitation and to foster efficient development of mineral resources.

2. A portion of these revenues shall be used, through or in cooperation with other international or regional organizations, to promote efficient, safe and economic exploitation of mineral resources of the seabed; to promote research on means to protect the marine environment; to advance other international efforts designed to promote safe and efficient use of the marine environment; to promote development of knowledge of the International Seabed Area; and to provide technical assistance to Contracting Parties or their nationals for these purposes, without discrimination.

ARTICLE 6

Neither this Convention nor any rights granted or exercised pursuant thereto shall affect the legal status of the superjacent waters as high seas, or that of the air space above those waters.

ARTICLE 7

All activities in the marine environment shall be conducted with reasonable regard for exploration and exploitation of the natural resources of the International Seabed Area.

ARTICLE 8

Exploration and exploitation of the natural resources of the International Seabed Area must not result in any unjustifiable interference with other activities in the marine environment.

ARTICLE 9

All activities in the International Seabed Area shall be conducted with strict and adequate safeguards for the protection of human life and safety and of the marine environment.

ARTICLE 10

All exploration and exploitation activities in the International Seabed Area shall be conducted by a Contracting Party or group of Contracting Parties or natural or juridicial persons under its or their authority or sponsorship.

ARTICLE 11

1. Each Contracting Party shall take appropriate measures to ensure that those conducting activities under its authority or sponsorship comply with this Convention.

2. Each Contracting Party shall make it an offense for those conducting activities under its authority or sponsorship in the International Seabed Area to violate the provisions of this Convention. Such offenses shall be punishable in accordance with administrative or judicial procedures established by the Authorizing or Sponsoring Party.

3. Each Contracting Party shall be responsible for maintaining public order on manned installations and equipment operated by those authorized or sponsored by it.

4. Each Contracting Party shall be responsible for damages caused by activities which it authorizes or sponsors to any other Contracting Party or its nationals.

5. A group of States acting together, pursuant to agreement among them or through an International organization, shall be jointly and severally responsible under this Convention.

ARTICLE 12

All disputes arising out of the interpretation or application of this Convention shall be settled in accordance with provisions of Section E of Chapter IV.

Chapter II—General Rules

A. Mineral Resources

ARTICLE 13

1. All exploration and exploitation of the mineral deposits of the International Seabed Area shall be licensed by the International Seabed Resource Authority or the appropriate Trustee Party. All licenses shall be subject to the provisions of this Convention.

2. Detailed rules to implement this Chapter are contained in Appendices A, B and C.

ARTICLE 14

1. There shall be fees for licenses for mineral exploration and exploitation.

2. The fees referred to in paragraph 1 shall be reasonable and be designed to defray the administrative expenses of the International Seabed Resources Authority and of the Contracting Parties in discharging their responsibilities in the International Seabed Area.

ARTICLE 15

1. An exploitation license shall specify the minerals or categories of minerals and the precise area to which it applies. The categories established

shall be those which will best promote simultaneous and efficient exploitation of different minerals.

2. Two or more licensees to whom licenses have been issued for different materials in the same or overlapping areas shall not unjustifiably interfere with each other's activities.

ARTICLE 16

The size of the area to which an exploitation license shall apply and the duration of the license shall not exceed the limits provided for in this Convention.

ARTICLE 17

Licensees must meet work requirements specified in this Convention as a condition of retaining an exploration license prior to and after commercial production is achieved.

ARTICLE 18

Licensees shall submit work plans and production plans, as well as reports and technical data acquired under an exploitation license, to the Trustee Party or the Sponsoring Party, as appropriate, and, to the extent specified by this Convention, to the International Seabed Resource Authority.

ARTICLE 19

1. Each Contracting Party shall be responsible for inspecting, at regular intervals, the activities of licensees authorized or sponsored by it. Inspection reports shall be submitted to the International Seabed Resource Authority.

2. International Seabed Authority, on its own initiative or at the request of any interested Contracting Party, may inspect any licensed activity in cooperation with the Trustee Party or Sponsoring Party, as appropriate, in order to ascertain that the licensed operation is being conducted in accordance with this Convention. In the event the International Seabed Resource Authority believes that a violation of this Convention has occurred, it shall inform the Trustee Party or Sponsoring Party, as appropriate, and request that suitable action be taken. If, after a reasonable period of time, the alleged violation continues, the International Seabed Resource Authority may bring the matter before the Tribunal in accordance with Section E of Chapter IV.

ARTICLE 20

1. Licenses issued pursuant to this Convention may be revoked only for cause in accordance with the provisions of this Convention.

2. Expropriation of investments made, or unjustifiable interference with operations conducted, pursuant to a license, is prohibited.

ARTICLE 21

1. Due notice must be given, by Notices to Mariners or other recognized means of notification, of the construction or deployment of any installations or devices for the exploration or exploitation of mineral deposits, and permanent means for giving warning of their presence must be maintained. Any installations or devices extending into the superjacent waters which are abandoned or disused must be entirely removed.

2. Such installations and devices shall not possess the status of islands and shall have no territorial sea of their own.

3. Installations or devices may not be established where interference with the use of recognized sea lanes or airways is likely to occur.

ARTICLE 22

Subject to the provisions of Chapter III, each Contracting Party may explore and exploit the seabed living resources of the International Seabed Area in accordance with such conservation measures as are necessary to protect the living resources of the International Seabed Area and to maximize their growth and utilization.

ARTICLE 23

1. In the International Seabed Area, the International Seabed Resource Authority shall prescribe Rules and Recommended Practices, in accordance with Chapter V of this Convention, to ensure:

    a. The protection of the marine environment against pollution arising from exploration and exploitation activities such as drilling, dredging, excavation, disposal of waste, construction and operation or maintenance of installations and pipelines and other devices;

    b. The prevention of injury to persons, property and marine resources arising from the aforementioned activities;

    c. The prevention of any unjustifiable interference with other activities in the marine environment arising from the aforementioned activities.

2. Deep drilling in the International Seabed Area shall be undertaken only in accordance with the provisions of this Convention.

D. Scientific Research

ARTICLE 24

1. Each Contracting Party agrees to encourage, and to obviate interference with, scientific research.

2. The Contracting Parties shall promote international cooperation in scientific research concerning the International Seabed Area:

a. By participating in international programs and by encouraging co-operation in scientific research by personnel of different countries;
b. Through effective publication of research programs and the results of research through international channels;
c. By cooperation in measures to strengthen the research capabilities of developing countries, including the participation of their nationals in research programs.

E. International Marine Parks and Preserves

ARTICLE 25

In consultation with the appropriate international organizations or agencies, the International Seabed Resource Authority may designate as international marine parks and preserves specific portions of the International Seabed Area that have unusual educational, scientific or recreational value. The establishment of such a park or preserve in the International Trusteeship Area shall require the approval of the appropriate Trustee Party.

Chapter III—The International Trusteeship

ARTICLE 26

1. The International Trusteeship Area is that part of the International Seabed Area comprising the continental or island margin between the boundary described in Article I and a line, beyond the base of the continental slope, or beyond the base of the slope of an island situated beyond the continental slope, where the downward inclination of the surface of the seabed declines to a gradient of 1 :————.

2. Each Trustee Party shall permanently delineate the precise seaward boundary of the International Trusteeship Area off its coast by straight lines not exceeding 60 nautical miles in length, following the general direction of the limits specified in paragraph 1. Such lines shall connect fixed points at the limit specified in paragraph 1, defined permanently by co-ordinates of latitude and longitude. Areas between or landward of such points may have a surface gradient of less than 1 :————. Where an elongate basin or plain having a surface gradient of less than 1 :———— transects an area having a gradient of more than 1 :————, a straight boundary line more than 60 nautical miles in length, but not exceeding the lesser of one-fourth of the length of that part of the basin or plain transecting the area having a gradient of more than 1 :———— or 120 nautical miles, may be drawn across the basin or plain.*

---

* The precise gradient should be determined by technical experts, taking into account, among other factors, ease of determination, the need to avoid dual administration of single mineral deposits, and the avoidance of including excessively large areas in the International Trusteeship Area.

3. Each Trustee Party shall submit the description of its boundary to the International Seabed Boundary Review Commission within five years of the entry into force of this Convention for that Party. Boundaries not accepted by that Commission and not resolved by negotiation between the Commission and the Trustee Party within one year shall be submitted by the Commission to the Tribunal for adjudication in accordance with Section E of Chapter IV.*

ARTICLE 27

1. Except as specifically provided for in this Chapter, the coastal State shall have no greater rights in the International Trusteeship Area off its coast than any other Contracting Party.

2. With respect to exploration and exploitation of the natural resources of that part of the International Trusteeship Area in which it acts as trustee for the international community, each coastal State, subject to the provisions of this Convention, shall be responsible for:

   a. Issuing, suspending and revoking mineral exploration and exploitation licenses;
   b. Establishing work requirements, provided that such requirements shall not be less than those specified in Appendix A;
   c. Ensuring that its licensees comply with this Convention, and, if it deems it necessary, applying standards to its licensees higher than or in addition to those required under this Convention, provided such standards are promptly communicated to the International Seabed Resource Authority;
   d. Supervising its licensees and their activities;
   e. Exercising civil and criminal jurisdiction over its licensees, and persons acting on their behalf, while engaged in exploration or exploitation;
   f. Filing reports with the International Seabed Resource Authority;
   g. Collecting and transferring to the International Seabed Resource Authority all payments required by this Convention;
   h. Determining the allowable catch of the living resources of the seabed and prescribing other conservation measures regarding them;
   i. Enacting such laws and regulations as are necessary to perform the above functions.

3. Detailed rules to implement this Chapter are contained in Appendix C.

ARTICLE 28

In performing the functions referred to in Article 27, the Trustee Party may, in its discretion:

   a. Establish the procedures for issuing licenses;

---

* Note.—Additional consideration will be given to problems raised by enclosed and semi-enclosed seas.

b. Decide whether a license shall be issued;

c. Decide to whom a license shall be issued, without regard to the provisions of Article 3;

d. Retain [a figure between 33⅓% and 50% will be inserted here] of all fees and payments required by this Convention;

e. Collect and retain additional license and rental fees to defray its administrative expenses, and collect, and retain [a figure between 33⅓% and 50% will be inserted here] of, other additional fees and payments related to the issuance or retention of a license, with annual notification to the International Seabed Resource Authority of the total amount collected;

f. Decide whether and by whom the living resources of the seabed shall be exploited, without regard to the provisions of Article 3.

ARTICLE 29

The Trustee Party may enter into an agreement with the International Seabed Resource Authority under which the International Seabed Resource Authority will perform some or all of the trusteeship supervisory and administrative functions provided for in this Chapter in return for an appropriate part of the Trustee Party's share of international fees and royalties.

ARTICLE 30

Where a part of the International Trusteeship Area is off the coast of two or more Contracting Parties, such Parties shall, by agreement, precisely delimit the boundary separating the areas in which they shall respectively perform their trusteeship functions and inform the International Seabed Boundary Review Commission of such delimitation. If agreement is not reached within three years after negotiations have commenced, the International Seabed Boundary Review Commission shall be requested to make recommendations to the Contracting Parties concerned regarding such delimitation. If agreement is not reached within one year after such recommendations are made, the delimitation recommended by the Commission shall take effect unless either Party, within 90 days thereafter, brings the matter before the Tribunal in accordance with Section E of Chapter IV.

Chapter IV—The International Seabed Resource Authority

A. General

ARTICLE 31

1. The International Seabed Resource Authority is hereby established.

2. The principal organs of the Authority shall be the Assembly, the Council, and the Tribunal.

ARTICLE 32

The permanent seat of the Authority shall be at —————.

ARTICLE 33

Each Contracting Party shall recognize the juridical personality of the Authority. The legal capacity, privileges and immunities of the Authority shall be the same as those defined in the Convention on the Privileges and Immunities of the Specialized Agencies of the United Nations.

B. The Assembly

ARTICLE 34

1. The Assembly shall be composed of all Contracting Parties.
2. The first session of the Assembly shall be convened —————. The Assembly shall thereafter be convened by the Council at least once every three years at a suitable time and place. Extraordinary sessions of the Assembly shall be convened at any time on the call of the Council, or the Secretary-General of the Authority at the request of one-fifth of the Contracting Parties.
3. At meetings of the Assembly a majority of the Contracting Parties is required to constitute a quorum.
4. In the Assembly each Contracting Party shall exercise one vote.
5. Decisions of the Assembly shall be taken by a majority of the members present and voting, except as otherwise provided in this Convention.

ARTICLE 35

The powers and duties of the Assembly shall be to:
a. Elect its President and other officers;
b. Elect members of the Council in accordance with Article 36;
c. Determine its rules of procedure and constitute such subsidiary organs as it considers necessary or desirable;
d. Require the submission of reports from the Council;
e. Take action on any matter referred to it by the Council;
f. Approve proposed budgets for the Authority, or return them to the Council for reconsideration and resubmission;
g. Approve proposals by the Council for changes in the allocation of the net income of the Authority within the limits prescribed in Appendix D, or return them to the Council for reconsideration and resubmission;
h. Consider any matter within the scope of this Convention and make recommendations to the Council or Contracting Parties as appropriate;

i. Delegate such of its powers as it deems necessary or desirable to the Council and revoke or modify such delegation at any time;

j. Consider proposals for amendments of this Convention in accordance with Article 76.

ARTICLE 36

1. The Council shall be composed of twenty-four Contracting Parties and shall meet as often as necessary.

2. Members of the Council shall be designated or elected in the following categories:

a. The six most industrially advanced Contracting Parties shall be designated in accordance with Appendix E;

b. Eighteen additional Contracting Parties, of which at least twelve shall be developing countries, shall be elected by the Assembly, taking into account the need for equitable geographical distribution.

3. At least two of the twenty-four members of the Council shall be landlocked or shelf-locked countries.

4. Elected members of the Council shall hold office for three years following the last day of the Assembly at which they are elected and thereafter until their successors are elected. Designated members of the Council shall hold office until replaced in accordance with Appendix E.

5. Representatives on the Council shall not be employees of the Authority.

ARTICLE 37

1. The Council shall elect its President for a term of three years.

2. The President of the Council may be a national of any Contracting Party, but may not serve during his term of office as its representative in the Assembly or on the Council.

3. The President shall have no vote.

4. The President shall:

a. Convene and conduct meetings of the Council;

b. Carry out the functions assigned to him by the Council.

ARTICLE 38

Decisions by the Council shall require approval by a majority of all its members, including a majority of members in each of the two categories referred to in paragraph 2 of Article 36.

ARTICLE 39

Any Contracting Party not represented on the Council may participate, without a vote, in the consideration by the Council or any of the subsidiary organs, of any question which is of particular interest to it.

ARTICLE 40

The powers and duties of the Council shall be to:

a. Submit annual reports to the Contracting Parties;

b. Carry out the duties specified in this Convention and any duties delegated to it by the Assembly;

c. Determine its rules of procedure;

d. Appoint and supervise the Commissions provided for in this Chapter, establish procedures for the coordination of their activities, and determine the terms of office of their members;

e. Establish other subsidiary organs, as may be necessary or desirable, and define their duties;

f. Appoint the Secretary-General of the Authority and establish general guidelines for the appointment of such other personnel as may be necessary;

g. Submit proposed budgets to the Assembly for its approval, and supervise their execution;

h. Submit proposals to the Assembly for changes in the allocation of the net income of the Authority within the limits prescribed in Appendix D;

i. Adopt and amend Rules and Recommended Practices in accordance with Chapter V, upon the recommendation of the Rules and Recommended Practices Commission;

j. Issue emergency orders, at the request of any Contracting Party, to prevent serious harm to the marine environment arising out of any exploration or exploitation activity and communicate them immediately to licensees, and Authorizing or Sponsoring Parties, as appropriate;

k. Establish a fund to provide emergency relief and assistance in the event of a disaster to the marine environment resulting from exploration or exploitation activities;

l. Establish procedures for coordination between the International Seabed Resource Authority, and the United Nations, its specialized agencies and other international or regional organizations concerned with the marine environment;

m. Establish or support such international or regional centers, through or in cooperation with other international and regional organizations, as may be appropriate to promote study and research of the natural resources of the seabed and to train nationals of any Contracting Party in related science and the technology of seabed exploration and exploitation, taking into account the special needs of developing States Parties to this Convention;

n. Authorize and approve agreements with a Trustee Party, pursuant to Article 29, under which the International Seabed Resource Authority will perform some or all of the Trustee Party's functions.

ARTICLE 41

In furtherance of Article 5, paragraph 2, of this Convention, the Council may, at the request of any Contracting Party and taking into account the special needs of developing States Parties to this Convention:

a. Provide technical assistance to any Contracting Party to further the objectives of this Convention;

b. Provide technical assistance to any Contracting Party to help it to meet its responsibilities and obligations under this Convention;

c. Assist any Contracting Party to augment its capability to derive maximum benefit from the efficient administration of the International Trusteeship Area.

ARTICLE 42

1. There shall be a Rules and Recommended Practices Commission, an Operations Commission, and an International Seabed Boundary Review Commission.

2. Each Commission shall be composed of five to nine members appointed by the Council from among persons nominated by Contracting Parties. The Council shall invite all Contracting Parties to submit nominations.

3. No two members of a Commission may be nationals of the same State.

4. A member of each Commission shall be elected its President by a majority of the members of the Commission.

5. Each Commission shall perform the functions specified in this Convention and such other functions as the Council may specify from time to time.

ARTICLE 43

1. Members of the Rules and Recommended Practices Commission shall have suitable qualifications and experience in seabed resources management, ocean sciences, maritime safety, ocean and marine engineering, and mining and mineral technology and practices. They shall not be full-time employees of the Authority.

2. The Rules and Recommended Practices Commission shall:

a. Consider, and recommend to the Council for adoption, Annexes to this Convention in accordance with Chapter V;

b. Collect from and communicate to Contracting Parties information which the Commission considers necessary and useful in carrying out its functions.

1. Members of the Operations Commission shall have suitable qualifications and experience in the management of seabed resources, and operation of marine installations, equipment and devices.

2. The Operations Commission shall:

a. Issue licenses for seabed mineral exploration and exploitation, except in the International Trusteeship Area;

b. Supervise the operations of licensees in cooperation with the Trustee or Sponsoring Party, as appropriate, but shall not itself engage in exploration or exploitation;

c. Perform such functions with respect to disputes between Contracting Parties as are specified in Section E of this Chapter;

d. Initiate proceedings pursuant to Section E of this Chapter for alleged violations of this Convention, including but not limited to proceedings for revocation or suspension of licenses;

e. Arrange for and review the collection of international fees and other forms of payment;

f. Arrange for the collection and dissemination of information relating to licensed operations;

g. Supervise the performance of the functions of the Authority pursuant to any agreement between a Trustee Party and the Authority under Article 29;

h. Issue deep drilling permits.

1. Members of the International Seabed Boundary Review Commission shall have suitable qualifications and experience in marine hydrography, bathymetry, geodesy and geology. They shall not be full-time employees of the Authority.

2. The International Seabed Boundary Review Commission shall:

a. Review the delineation of boundaries submitted by Contracting Parties in accordance with Articles 1 and 26 to see that they conform to the provisions of this Convention, negotiate any differences with Contracting Parties, and if these differences are not resolved initiate proceedings before the Tribunal in accordance with Section E of this Chapter;

b. Make recommendations to the Contracting Parties in accordance with Article 30;

c. At the request of any Contracting Party, render advice on any boundary question arising under this Convention.

E. The Tribunal

1. The Tribunal shall decide all disputes and advise on all questions relating to the interpretation and application of this Convention which have

been submitted to it in accordance with the provisions of this Convention. In its decisions and advisory opinions the Tribunal shall also apply relevant principles of international law.

2. Subject to an authorization under Article 96 of the Charter of the United Nations, the Tribunal may request the International Court of Justice to give an advisory opinion on any question of international law.

ARTICLE 47

1. The Tribunal shall be composed of five, seven, or nine independent judges, who shall possess the qualifications required in their respective countries for appointment to the highest judicial offices, or shall be lawyers especially competent in matters within the scope of this Convention. In the Tribunal as a whole the representation of the principal legal systems of the world shall be assured.

2. No two of the members of the Tribunal may be nationals of the same State.

ARTICLE 48

1. Each Contracting Party shall be entitled to nominate candidates for membership on the Tribunal. The Council shall elect the Tribunal from a list of these nominations.

2. The members of the Tribunal shall be elected for nine years and may be re-elected, providing, however, that the Council may establish procedures for staggered terms. Should such procedures be established, the judges whose terms are to expire in less than nine years shall be chosen by lots drawn by the Secretary-General.

3. The members of the Tribunal shall continue to discharge their duties until their places have been filled. Though replaced, they shall finish any cases which they may have begun.

4. A member of the Tribunal unable to perform his duties may be dismissed by the Council on the unanimous recommendation of the other members of the Tribunal.

5. In case of a vacancy, the Council shall elect a successor who shall hold office for the remainder of his predecessor's term.

ARTICLE 49

The Tribunal shall establish its rules of procedure; elect its President; appoint its Registrar and determine his duties and terms of service; and adopt regulations for the appointment of the remainder of its staff.

ARTICLE 50

1. Any Contracting Party which considers that another Contracting Party has failed to fulfill any of its obligations under this Convention may bring its complaint before the Tribunal.

2. Before a Contracting Party institutes such proceedings before the Tribunal it shall bring the matter before the Operations Commission.

3. The Operations Commission shall deliver a reasoned opinion in writing after the Contracting Parties concerned have been given the opportunity both to submit their own cases and to reply to each other's case.

4. If the Contracting Party accused of a violation does not comply with the terms of such opinion within the period laid down by the Commission, the other Party concerned may bring the matter before the Tribunal.

5. If the Commission has not given an opinion within a period of three months from the date when the matter was brought before it, either Party concerned may bring the matter before the Tribunal without waiting further for the opinion of the Commission.

ARTICLE 51

1. Whenever the Operations Commission, acting on its own initiative or at the request of any licensee, considers that a Contracting Party or a licensee has failed to fulfill any of its obligations under this Convention, it shall issue a reasoned opinion in writing on the matter after giving such party the opportunity to submit its comments.

2. If the party concerned does not comply with the terms of such opinion within the period laid down by the Commission, the latter may bring a complaint before the Tribunal.

ARTICLE 52

1. If the Tribunal finds that Contracting Party or a licensee has failed to fulfill any of its obligations under this Convention, such party shall take the measures required for the implementation of the judgment of the Tribunal.

2. When appropriate, the Tribunal may decide that the Contracting Party or the licensee who has failed to fulfill its obligations under this Convention shall pay to the Authority a fine of not more that $1,000 for each day of the offense, or shall pay damages to the other party concerned, or both.

3. In the event the Tribunal determines that a licensee has committed a gross and persistent violation of the provisions of this Convention and has not within a reasonable time brought his operations into compliance with them, the Council may, as appropriate, either revoke his license or request that the Trustee Party revoke it. The licensee shall not, however, be deprived of his license if his actions were directed by a Trustee or Sponsoring Party.

ARTICLE 53

If disputes under Articles 1, 26 and 30 have not been resolved by the time and methods specified in those Articles, the International Seabed Boundary Review Commission shall bring the matter before the Tribunal.

1. Any Contracting Party, which questions the legality of measures taken by the Council, the Rules and Recommended Practices Commission, the Operations Commission, or the International Seabed Boundary Review Commission on the grounds of a violation of this Convention, lack of jurisdiction, infringement of important procedural rules, unreasonableness, or misuse of powers, may bring the matter before the Tribunal.

2. Any person may, subject to the same conditions, bring a complaint to the Tribunal with regard to a decision directed to that person, or a decision which, although in the form of a rule or a decision directed to another person, is of direct concern to the Complainant.

3. The proceedings provided for in this Article shall be instituted within a period of two months, dating, as the case may be, either from the publication of the measure concerned or from its notification to the complainant, or, in default thereof, from the day on which the latter learned of it.

4. If the Tribunal considers the appeal well-founded, it should declare the measure concerned to be null and void, and shall decide to what extent the annulment shall have retroactive application.

ARTICLE 55

1. The organ responsible for a measure declared null and void by the Tribunal shall be required to take the necessary steps to comply with the Tribunal's judgment.

2. When appropriate, the Tribunal may require that the Authority repair or pay for any damage caused by its organs or by its officials in the performance of their duties.

ARTICLE 56

When a case pending before a court or tribunal of one of the Contracting Parties raises a question of the interpretation of this Convention or of the validity or interpretation of measures taken by an organ of the Authority, the court or tribunal concerned may request the Tribunal to give its advice thereon.

ARTICLE 57

The Tribunal shall also be competent to decide any dispute connected with the subject matter of this Convention submitted to it pursuant to an agreement, license, or contract.

ARTICLE 58

If a Contracting Party fails to perform the obligations incumbent upon it under a judgment rendered by the Tribunal, the other Party to the case may have recourse to the Council, which shall decide upon measures to be taken to give effect to the judgment. When appropriate, the Council may decide to suspend temporarily, in whole or in part, the rights under this Convention of the Party failing to perform its obligations, without impairing the rights of licensees who have not contributed to the failure to perform such obligations. The extent of such a suspension should be related to the extent and seriousness of the violation.

ARTICLE 59

In any case in which the Council issues an order in emergency circumstances to prevent serious harm to the marine environment, any directly affected Contracting Party may request immediate review by the Tribunal, which shall promptly either confirm or suspend the application of the emergency order pending the decision of the case.

ARTICLE 60

Any organ of the International Seabed Resource Authority may request the Tribunal to give an advisory opinion on any legal question connected with the subject matter of this Convention.

F. The Secretariat

ARTICLE 61

The Secretariat shall comprise a Secretary-General and such staff as the International Seabed Resources Authority may require. The Secretary-General shall be appointed by the Council from among persons nominated by Contracting Parties. He shall serve for a term of six years, and may be reappointed.

ARTICLE 62

The Secretary-General shall:
a. Be the chief administrative officer of the International Seabed Resource Authority, and act in that capacity in all meetings of the Assembly and the Council;
b. Report to the Assembly and the Council on the work of the International Seabed Resource Authority;
c. Collect, publish and disseminate information which will contribute to mankind's knowledge of the seabed and its resources;
d. Perform such other functions as are entrusted to him by the Assembly or the Council.

ARTICLE 63

1. In the performance of their duties the Secretary-General and the staff shall not seek or receive instructions from any government or from any other external authority. They shall refrain from any action which

might reflect on their position as international officials responsible only to the International Seabed Resource Authority.

2. Each Contracting Party shall respect the exclusively international character of the responsibilities of the Secretary-General and the staff and shall not seek to influence them in the discharge of their responsibilities.

ARTICLE 64

1. The staff of the International Seabed Resource Authority shall be appointed by the Secretary-General under the general guidelines established by the Council.

2. Appropriate staffs shall be assigned to the various organs of the Authority as required.

3. The paramount consideration in the employment of the staff and in the determination of the conditions of service shall be the necessity of securing the highest standards of efficiency, competence, and integrity. Due regard shall be paid to the importance of recruiting the staff on as wide a geographical basis as possible.

G. Conflicts of Interest

ARTICLE 65

No representative to the Assembly or the Council nor any member of the Tribunal, Commission, subsidiary organs (other than advisory bodies or consultants), or the Secretariat, shall, while serving as such a representative or member, be actively associated with or financially interested in any of the operations of any enterprise concerned with exploration or exploitation of the natural resources of the International Seabed Area.

Chapter V—Rules and Recommended Practices

ARTICLE 66

1. Rules and Recommended Practices are contained in Annexes to this Convention.

2. Annexes shall be consistent with this Convention, its Appendices, and any amendments thereto. Any Contracting Party may challenge an Annex, an amendment to an Annex, or any of their provisions, on the grounds that it is unnecessary, unreasonable or constitutes a misuse of powers, by bringing the matter before the Tribunal in accordance with Article 54.

3. Annexes shall be adopted and amended in accordance with Article 67. Those Annexes adopted along with this Convention, if any, may be amended in accordance with Article 67.

ARTICLE 67

The Annexes to this Convention and amendments to such Annexes shall be adopted in accordance with the following procedure:

a. They shall be prepared by the Rules and Recommended Practices Commission and submitted to the Contracting Parties for comments;

b. After receiving the comments, the Commission shall prepare a revised text of the Annex or amendments thereto;

c. The text shall then be submitted to the Council which shall adopt it or return it to the Commission for further study;

d. If the Council adopts the text, it shall submit it to the Contracting Parties;

e. The Annex or an amendment thereto shall become effective within three months after its submission to the Contracting Parties, or at the end of such longer period of time as the Council may prescribe, unless in the meantime more than one-third of the Contracting Parties register their disapproval with the Authority;

f. The Secretary-General shall immediately notify all Contracting States of the coming into force of any Annex or amendment thereto.

ARTICLE 68

1. Annexes shall be limited to the Rules and Recommended Practices necessary to:

a. Fix the level, basis, and accounting procedures for determining international fees and other forms of payment, within the ranges specified in Appendix A;

b. Establish work requirements within the ranges specified in Appendices A and B;

c. Establish criteria for defining the technical and financial competence of applicants for licenses;

d. Assure that all exploration and exploitation activities, and all deep drilling, are conducted with strict and adequate safeguards for the protection of human life and safety and of the marine environment;

e. Protect living marine organisms from damage arising from exploration and exploitation activities;

f. Prevent or reduce to acceptable limits interference arising from exploration and exploitation activities with other uses and users of the marine environment;

g. Assure safe design and construction of fixed exploration and exploitation installations and equipment;

h. Facilitate search and rescue services, including assistance to aquanauts, and the reporting of accidents;

i. Prevent unnecessary waste in the extraction of minerals from the seabed;

j. Standardize the measurement of water depth and the definition of other natural features pertinent to the determination of the precise location of International Seabed Area boundaries;

k. Prescribe the form in which Contracting Parties shall describe their boundaries and the kinds of information to be submitted in support of them;

l. Encourage uniformity in seabed mapping and charting;

m. Facilitate the management of a part of the international trusteeship area pursuant to any agreement between a Trustee Party and the Authority under Article 29;

n. Establish and prescribe conditions for the use of international marine parks and preserves;

2. Application of any Rule or Recommended Practice may be limited as to duration or geographic area, but without discrimination against any Contracting Party or licensee.

ARTICLE 69

The Contracting Parties agree to collaborate with each other and the appropriate Commission in securing the highest practicable degree of uniformity in regulations, standards, procedures and organizations in relation to the matters covered by Article 68 in order to facilitate and improve seabed resources exploration and exploitation.

ARTICLE 70

Annexes and amendments thereto shall take into account existing international agreements and, where appropriate, shall be prepared in collaboration with other competent international organizations. In particular, existing international agreements and regulations relating to safety of life at sea shall be respected.

ARTICLE 71

1. Except as otherwise provided in this Convention, the Annexes and amendments thereto adopted by the Council shall be binding on all Contracting Parties.

2. Recommended Practices shall have no binding effect.

ARTICLE 72

Any Contracting Party believing that a provision of an Annex or an amendment thereto cannot be reasonably applied because of special circumstances may seek a waiver from the Operations Commission and if such waiver is not granted within three months, it may appeal to the Tribunal within an additional period of two months.

Chapter VI—Transition

ARTICLE 73

1. There shall be due protection for the integrity of investments made in the International Seabed Area prior to the coming into force of this Convention.

2. All authorizations by a Contracting Party to exploit the mineral resources of the International Seabed Area granted prior to July 1, 1970, shall be continued without change after the coming into force of this Convention provided that:

    a. Activities pursuant to such authorizations shall, to the extent possible, be conducted in accordance with the provisions of this Convention;

    b. New activities under such previous authorizations which are begun after the coming into force of this Convention shall be subject to the regulatory provisions of this Convention regarding the protection of human life and safety and of the marine environment and the avoidance of unjustifiable interference with other uses of the marine environment;

    c. Upon the expiration or relinquishment of such authorizations, or upon their revocation by the authorizing Party, the provisions of this Convention shall become fully applicable to any exploration or exploitation of resources remaining in the areas included in such authorizations;

    d. Contracting Parties shall pay to the International Seabed Resource Authority, with respect to such authorizations, the production payments provided for under this Convention.

3. A Contracting Party which has authorized exploitation of the mineral resources of the International Seabed Area on or after July 1, 1970, shall be bound, at the request of the person so authorized, either to issue new licenses under this Convention in its capacity as a Trustee Party, or to sponsor the application of the person so authorized to receive new licenses from the International Seabed Resource Authority. Such new license issued by a Trustee Party shall include the same terms and conditions as its previous authorization, provided that such license shall not be inconsistent with this Convention, and provided further that the Trustee Party shall itself be responsible for complying with increased obligations resulting from the application of this Convention, including fees and other payments required by this Convention.

4. The provisions of paragraph 3 shall apply within one year after this Convention enters into force for the Contracting Party concerned, but in no event more than five years after the entry into force of this Convention.

5. Until converted into new licenses under paragraph 3, all authoriza-

tions issued on or after July 1, 1970, to exploit the mineral resources of the International Seabed Area shall have the same status as authorizations under paragraph 2. Five years after the entry into force of this Convention all such authorizations not converted into new licenses under paragraph 3 shall be null and void.

6. Any Contracting Party that has authorized activities within the International Seabed Area after July 1, 1970, but before this Convention has entered into force for such Party, shall compensate its licensees for any investment losses resulting from the application of this Convention.

ARTICLE 74

1. The membership of the Tribunal, the Commissions, and the Secretariat shall be maintained at a level commensurate with the tasks being performed.

2. In the period before the International Seabed Resource Authority acquires income sufficient for the payment of its administrative expenses, the Authority may borrow funds for the payment of those expenses. The Contracting Parties agree to give sympathetic consideration to requests by the Authority for such loans.

Chapter VII—Definitions

ARTICLE 75

Unless another meaning results from the context of a particular provision, the following definitions shall apply:

1. "Convention" refers to all provisions of and amendments to this Convention, its Appendices, and its Annexes.

2. "Trustee Party" refers to the Contracting Party exercising trusteeship functions in that part of the International Trusteeship Area off its coast in accordance with Chapter III.

3. "Sponsoring Party" refers to a Contracting Party which sponsors an application for a license or permit before the International Seabed Resource Authority. The term "sponsor" is used in this context.

4. "Authorizing Party" refers to a Contracting Party authorizing any activity in the International Seabed Area, including a Trustee Party issuing exploration or exploitation licenses. The term "authorize" is used in this context. In the case of a vessel, the term "Authorizing Party" shall be deemed to refer to the State of its nationality.

5. "Operating Party" refers to a Contracting Party which itself explores or exploits the natural resources of the International Seabed Area.

6. "Licensee" refers to a State, group of States, or natural or juridical person holding a license for exploration or exploitation of the natural resources of the International Seabed Area.

7. "Exploration" refers to any operation in the International Seabed Area which has as its principal or ultimate purpose the discovery and appraisal, or exploitation, of mineral deposits, and does not refer to scientific research. The term does not refer to similar activities when undertaken pursuant to an exploitation license.

8. "Deep drilling" refers to any form of drilling or excavation in the International Seabed Area deeper than 300 meters below the surface of the seabed.

9. "Landlocked or shelf-locked country" refers to a Contracting Party which is not a Trustee Party.

Chapter VIII—Amendment and Withdrawal

ARTICLE 76

Any proposed amendment to this Convention or the appendices thereto which has been approved by the Council and a two-thirds vote of the Assembly shall be submitted by the Secretary-General to the Contracting Parties for ratification in accordance with their respective constitutional processes. It shall come into force when ratified by two-thirds of the Contracting Parties, including each of the six States designated pursuant to subparagraph 2(a) of Article 36 at the time the Council approved the amendments. Amendments shall not apply retroactively.

ARTICLE 77

1. Any Contracting Party may withdraw from this Convention by a written notification addressed to the Secretary-General. The Secretary-General shall promptly inform the other Contracting Parties of any such withdrawal.

2. The withdrawal shall take effect one year from the date of the receipt by the Secretary-General of the notification.

Chapter IX—Final Clauses

ARTICLE 78

Appendix A:

Terms and Procedures Applying to All Licenses in the International Seabed Area

1. Activities Requiring a License or a Permit:

1.1. Pursuant to Article 13 of this Convention, all exploration and exploitation operations in the International Seabed Area which have as their principal or ultimate purpose the discovery or appraisal, and exploitation, of mineral deposits shall be licensed.

1.2. There shall be two categories of licenses:

a. A non-exclusive exploration license shall authorize geophysical and geochemical measurements, and bottom sampling, for the purposes of exploration. This license shall not be restricted as to area and shall grant no exclusive right to exploration nor any preferential right in applying for an exploitation license. It shall be valid for two years following the date of its issuance and shall be renewable for successive two-year periods.

b. An exploitation license shall authorize exploration and exploitation of one of the groups of minerals described in Section 5 of this Appendix in a specified area. The exploitation license shall include the exclusive right to undertake deep drilling and other forms of subsurface entry for the purpose of exploration and exploitation of minerals described in paragraphs 5.1(a) and 5.1(c). The license shall be for a limited period and shall expire at the end of fifteen years if no commercial production is achieved.

1.3. The right to undertake deep drilling for exploration or exploitation shall be granted only under an exploitation license.

1.4. Deep drilling for purposes other than exploration or exploitation of seabed minerals shall be authorized under a deep-drilling permit issued at no charge by the International Seabed Resource Authority, provided that:

a. The application is accompanied by a statement from the Sponsoring Party certifying as to the applicant's technical competence and accepting liability for any damages that may result from such drilling;

b. The application for such a permit is accompanied by a description of the location proposed for such holes, by seismograms and other pertinent information on the geology in the vicinity of the proposed drilling sites, and by a description of the equipment and procedures to be utilized;

c. The proposed drilling, including the methods and equipment to be utilized, complies with the requirements of this Convention and is judged by the Authority not to pose an uncontrollable hazard to human safety, property, and the environment;

d. The proposed drilling is either not within an area already under an exploitation license or is not objected to by the holder of such a license;

e. The applicant agrees to make available promptly the geologic information obtained from such drilling to the Authority and the public.

2. General License Procedures:

2.1. An Authorizing or Sponsoring Party shall certify the operator's financial and technical competence and shall require the operator to conform to the rules, provisions and procedures specified under the terms of the license.

2.2. Each Authorizing or Sponsoring Party shall formulate procedures to ensure that applications for licenses are handled expeditiously and fairly.

2.3. Any Authorizing or Sponsoring Party which considers that it is unable to exercise appropriate supervision over operators authorized or sponsored by it in accordance with this Convention shall be permitted to authorize or sponsor operators only if their operations are supervised by the International Seabed Resource Authority pursuant to an agreement between the Authorizing or Sponsoring Party and the International Seabed Resource Authority. In such event fees and rentals normally payable to the International Seabed Resource Authority will be increased appropriately to offset its supervisory costs.

3. Exploration Licenses—Procedures:

3.1. All applications for exploration licenses and for their renewal shall be accompanied by a fee of from $500 to $1,500 as specified in an Annex and a description of the location of the general area to be investigated and the kinds of activities to be undertaken. A portion [a figure between 50% and 66⅔% will be inserted here] of the fee shall be forwarded by the Authorizing or Sponsoring Party to the Authority together with a copy of the application.

3.2. The Authorizing or Sponsoring Party shall transmit to the Authority the description referred to in paragraph 3.1 and its assurance that the activities will not be harmful to the marine environment.

3.3. The Authorizing or Sponsoring Party may require the operator to pay, and may retain, an additional license fee not to exceed $3,000, to help cover the administrative expenses of that Party.

3.4. Exploration licenses shall not be renewed in the event the operator has failed to conform his activities under the prior license to the provisions of this Convention or to the conditions of the license.

4. Exploitation Licenses—Procedures:

4.1. All applications for exploitation licenses shall be accompanied by a fee of from $5,000 to $15,000, per block, as specified in an Annex. A portion [a figure between 50% and 66⅔% will be inserted here] of the fee shall be forwarded by the Authorizing or Sponsoring Party to the Authority together with a copy of the application.

4.2. Pursuant to Section 5 of this Appendix, applications shall identify the category of minerals in the specific area for which a license is sought.

4.3. When a license is granted to an applicant for more than one block at the same time, only a single certificate need be issued.

4.4. The Authorizing or Sponsoring Party may require the operator to pay, and may retain, an additional license fee not to exceed $30,000, to help cover the administrative expenses of that Party.

4.5. The license fee described in paragraph 4.1 shall satisfy the first two years' rental fee.

5. Exploitation Rights—Categories and Size of Blocks:

5.1. Licenses to exploit shall be limited to one of the following categories of minerals:

    a. Fluids or minerals extracted in a fluid state, such as oil, gas, helium, nitrogen, carbon dioxide, water, geothermal energy, sulfur and saline minerals.

    b. Manganese-oxide nodules and other minerals at the surface of the seabed.

    c. Other minerals, including category (b) minerals that occur beneath the surface of the seabed and metalliferous muds.

5.2. An exploitation license shall be issued for a specific area of the seabed and subsoil vertically below it, hereinafter referred to as a "block." The methods for defining the boundaries of blocks, and of portions thereof, shall be specified in an Annex.

5.3. In the category described in paragraph 5.1(a) the block shall be approximately 500 square kilometers, which shall be reduced to a quarter of a block when production begins. Each exploitation license shall apply to not more than one block, but exploitation licenses to a rectangle containing as many as 16 contiguous blocks may be taken out under a single certificate and reduced by three quarters to a number of blocks, a single block, or a portion of a single block when production begins. The relinquishment requirement shall not apply to licenses issued for areas of one quarter of a block or less.

5.4. In the category described in paragraph 5.1(b) the block shall be approximately 40,000 square kilometers, which shall be reduced to a quarter of a block when production begins. Each exploitation license shall apply to not more than one block, but exploitation licenses to a rectangle containing as many as four contiguous blocks may be taken out under a single certificate and reduced to a single block, or to a portion of a single block, comprising one-fourth their total area, when production begins. The relinquishment requirement shall not apply to licenses issued for areas of one quarter of a block or less.

5.5. In the category described in paragraph 5.1(c) the block shall be approximately 500 square kilometers, which shall be reduced to one eighth of a block when production begins. Each license shall apply to not more than one block, but exploitation licenses to as many as 8 contiguous blocks may be taken out under a single certificate and reduced to a single block, or to a portion of a single block, comprising one eighth their total area, when production begins. The relinquishment shall not apply to licenses issued for one eighth of a block or less.

5.6. Applications for exploitation licenses may be for areas smaller than the maximum stated above.

5.7. Operators may at any time relinquish rights to all or part of the licensed area.

5.8. Commercial production shall be deemed to have commenced or to be maintained when the value at the site of minerals exploited is not less than $100,000 per annum. The required minimum and the method of ascertaining this value shall be determined by the Authority.

5.9. If the commercial production is not maintained, the exploitation license shall expire within five years of its cessation, but when production is interrupted or suspended for reasons beyond the operator's control, the duration of the license shall be extended by a time equal to the period in which production has been suspended for reasons beyond the operator's control.

6. Rental Fees and Work Requirements:

RENTAL FEES

6.1. Prior to attaining commercial production the following annual rental fees shall be paid beginning in the third year after the license has been issued: (a) $2–$10 per square kilometer, as specified in an appropriate Annex, for the category of minerals described in paragraph 5.1(a) above; (b) $2–$10 per 100 square kilometers for the category of minerals described in paragraph 5.1(b) above, and (c) $2–$10 per square kilometer for the category of minerals described in paragraph 5.1(c) above.

6.2. The rates in paragraph 6.1 shall increase at the rate of 10% per annum, calculated on the original base rental fee, for the first ten years after the third year, and shall increase 20% per annum, calculated on the original base rental fee, for the following two years.

6.3. After commercial production begins, the annual rental fee shall be $5,000–$25,000 per block, regardless of block size.

6.4. The rental fee shall be payable annually in advance to the Authorizing or Sponsoring Party which shall forward a portion [a figure between 50% and 66⅔% will be inserted here] of the fees to the Authority. The Authorizing or Sponsoring Party may require the operator to pay, and may retain, an additional rental fee, not to exceed an amount equal to the amount paid pursuant to paragraphs 6.1 through 6.3, to help cover the administration expenses of that Party.

WORK REQUIREMENTS

6.5. Prior to attaining commercial production, the operator shall deposit a work requirement fee, or post a sufficient bond for that amount, for each license at the beginning of each year.

6.6. The minimum annual work requirement fee for each block shall increase in accordance with the following schedule:

Paragraph 5.1 (a) and (c) minerals

Years:	Amounts per annum
1 to 5	$20,000
6 to 10	180,000
11 to 15	200,000
Total	2,000,000

Paragraph 5.1 (b) minerals

Years:	Amounts per annum
1 to 2	$20,000
3 to 10	120,000
11 to 15	200,000
Total	2,000,000

The minimum annual work requirement fee for a portion of a block shall be an appropriate fraction of the above, to be specified in an Annex.

6.7. The work requirement fee shall be refunded to the operator upon receipt of proof by the Authorizing Party or Sponsoring Party that the amount equivalent to the fee has been expended in actual operations. Expenditures for on-land design or process research and equipment purchase or off-site construction cost directly related to the licensed block or group of blocks shall be considered to apply toward work requirements up to 75% of the amount required.

6.8. Expenditures in excess of the required amount for any given year shall be credited to the requirement for the subsequent year or years.

6.9. In the absence of satisfactory proof that the required expenditure has been made in accordance with the foregoing provisions of this section, the deposit will be forfeited.

6.10. If cumulative work requirement expenditures are not met at the end of the initial five-year period, the exploitation license shall be forfeited.

6.11. After commercial production begins the operator shall make an annual deposit of at least $100,000 at the beginning of each year; or shall post a sufficient bond for that amount, which shall be refunded in an amount equivalent to expenditures on or related to the block and the value of production at the site.

6.12. If production is suspended or delayed for reasons beyond the operator's control, the operator shall not be required to make the deposit or post the bond required in paragraph 6.11.

7. Submission of Work Plans and Data Under Exploitation Licenses Prior to Commencement of Commercial Production:

7.1. Exploitation license applications shall be accompanied by a general description of the work to be done and the equipment and methods to be used. The licensee shall submit subsequent changes in his work plan to the Sponsoring or Authorizing Party for review.

7.2. The licensee shall furnish reports at specific intervals to the Authorizing or Sponsoring Party supplying proof that he has fulfilled the specified work requirements. Copies of such reports shall be forwarded to the Authority.

7.3. The licensee shall maintain records of drill logs, geophysical data and other data acquired in the area to which his license refers, and shall provide access to them to the Authorizing or Sponsoring Party on request.

7.4. At intervals of five years, or when he relinquishes his rights to all or part of the area or when he submits a production plan as described in Section 8 of this Appendix, the operator shall transmit to the Authorizing or Sponsoring Party such maps, seismic sections, logs, assays, or reports, as are specified in an Annex to this Convention. The Authorizing or Sponsoring Party shall hold such data in confidence for ten years after receipt, but shall make the data available on request to the Authority for its confidential use in the inspection of operations.

7.5. The data referred to in paragraph 7.4 shall be transmitted to the Authority ten years after receipt by the Authorizing or Sponsoring Party, and made available by the Authority for public inspection. Such data shall be transmitted to the Authority immediately upon revocation of a license.

8. Production Plan and Producing Operations:

8.1. Prior to beginning commercial production the licensee shall submit a production plan to the Authorizing or Sponsoring Party and through such Party to the Authority.

8.2. The Authorizing or Sponsoring Party and the Authority shall require such modifications in the plan as may be necessary for it to meet the requirements of this Convention.

8.3. Any change in the licensee's production plan shall be submitted to the Authorizing or Sponsoring Party and through such Party to the Authority for their review and approval.

8.4. Not later than three months after the end of each year from the issuance of the license the licensee shall transmit to the Authorizing or Sponsoring Party for forwarding to the Authority production reports and such other data as may be specified in an Annex to this Convention.

8.5. The operator shall maintain geologic, geophysical and engineering records and shall provide access to them to the Authorizing or Sponsoring Party on its request. In addition, the operator shall submit annually such maps, sections, and summary reports, as are specified in Annexes to this Convention.

8.6. The Sponsoring or Authorizing Party shall hold such maps and reports in confidence for ten years from the time received but shall make them available on request to the Authority for its confidential use in the inspection of operations.

8.7. Such maps and reports shall be transmitted to the Authority and shall be made available by it for public inspection not later than ten years after receipt by the Sponsoring or Authorizing Party.

9. Unit Operations:

9.1. Accumulations of fluids and other minerals that can be made to migrate from one block to another and that would be most rationally mined by an operation under the control of a single operator but that lie astride the boundary of adjacent blocks licensed to different operators shall be brought into unit management and production.

9.2. With respect to deposits lying astride the seaward boundary of the International Trusteeship Area, the Operations Commission shall assure unit management and production, giving the Trustee and Sponsoring Parties and their licensees a reasonable time to reach agreement on an operation plan.

10. Payments on Production:

10.1. When commercial production begins under an exploitation license, the operator shall pay a cash production bonus of $500,000 to $2,000,000 per block, as specified in an Annex to this Convention, to the Authorizing or Sponsoring Party.

10.2. Thereafter, the operator shall make payments to the Authorizing or Sponsoring Party which are proportional to production, in the nature of total payments ordinarily made to governments under similar conditions. Such payments shall be equivalent to 5 to 40 percent of the gross value at the site of oil and gas, and 2 to 20 percent of the gross value at the site of other minerals, as specified in an Annex to this Convention. The total annual payment shall not be less than the annual rental fee under paragraph 6.3.

10.3. The Sponsoring Party shall forward all payments under this section to the Authority. The Authorizing Party shall forward a portion [a figure between 50% and 66⅔% will be inserted here] of such payments to the Authority.

11. Graduation of Payments According to Environment and Other Factors:

11.1. The levels of payments and work requirements, as well as the rates at which such payments and work requirements escalate over time, may be graduated to take account of probable risk and cost to the investor, including such factors as water depth, climate, volume of production, proximity to existing production, or other factors affecting the economic rent that can reasonably be anticipated from mineral production in a given area.

11.2. Any graduated levels and rates shall be described and categorized in an Annex in such a way as to affect all licensees in each category equally and not to discriminate against or favor individual Parties or groups of Parties, or their nationals.

11.3. Any increases in such levels of payment or requirements shall apply only to new licenses or renewals and not to those already in force.

12. Liability:

12.1. The operator and his Authorizing or Sponsoring Party, as appropriate, shall be liable for damage to other users of the marine environment and for clean-up and restoration costs of damage to the land environment.

12.2. The Authorizing or Sponsoring Party, as appropriate, shall require operators to subscribe to an insurance plan or provide other means of guaranteeing responsibility, adequate to cover the liability described in paragraph————.*

13. Revocation:

13.1. In the event of revocation pursuant to Article 52 of this Convention, there shall be no reimbursement for any expense incurred by the licensee prior to the revocation. The licensee shall, however, have the right to recover installations or equipment within six months of the date of the revocation of his license. Any installations or devices not removed by that time shall be removed and disposed of by the Authority, or the Authorizing or Sponsoring Party, at the expense of the licensee.

* NOTE.—More detailed provisions on liability should be included.

14. International Fees and Payments:

14.1. The Authority shall specify the intervals at which fees and other payments collected by an Authorizing or Sponsoring Party shall be transmitted.

14.2. No Contracting Party shall impose or collect any tax, direct or indirect, on fees and other payments to the Authority.

14.3. All fees and payments required under this Convention shall be those in force at the time a license was issued or renewed.

14.4. All fees and payments to the Authority shall be transmitted in convertible currency.

Appendix B:

Terms and Procedures Applying to Licenses in the International Seabed Area Beyond the International Trusteeship Area

1. Entities Entitled to Obtain Licenses:

1.1. Contracting Parties or a group of Contracting Parties, one of which shall act as the operating or sponsoring Party for purposes of fixing operational or supervisory responsibility, are authorized to apply for and obtain exploration and exploitation licenses. Any Contracting Party or group of Contracting Parties, which applies for a license to engage directly in exploration or exploitation, shall designate a specific agency to act as operator on its behalf for the purposes of this Convention.

1.2. Natural or juridical persons are authorized to apply for and obtain exploration and exploitation licenses from the International Seabed Resource Authority if they are sponsored by a Contracting Party.

2. Exploration Licenses—Procedures:

2.1. Licenses shall be issued promptly by the Authority through the Sponsoring Party to applicants meeting the requirements specified in . Appendix A.

3. Exploitation Licenses—Procedures:

3.1. The Sponsoring Party shall certify as to the technical and financial competence of the operator, and shall transmit the operator's work plan.

3.2. An application for an exploitation license shall be preceded by a notice of intent to apply for a license submitted by the operator to the Authority and the prospective Sponsoring Party. Such a notice of intent, when accompanied by evidence of the deposit of the license fee referred to in paragraph 4.1 of Appendix A, shall reserve the block for one hundred and eighty days. Notices of intent may not be renewed.

3.3. Notices of intent shall be submitted sealed to the Authority and opened at monthly intervals at previously announced times.

3.4. Subject to compliance with these procedures, if only one notice of intent has been received for a particular block, the applicant shall be granted a license, except as provided in paragraphs 3.6 through 3.8.

3.5. If more than one notice of intent to apply for a license for the same block or portion thereof is received at the same opening, the Authority shall notify the applicants and their Sponsoring Parties that the exploitation license to the block or portion thereof will be sold to the highest bidder at a sale to be held one hundred and eighty days later, under the following terms:

a. The bidding shall be on a cash bonus basis and the minimum bid shall be twice the license fee;

b. Bids shall be sealed;

c. The bidding shall be limited to such of the original applicants whose applications have been received in the interim from their sponsoring Parties;

d. Bids shall be announced publicly by the Authority when they are opened. In the event of a tie, the tie bidders shall submit a second sealed bid to be opened 28 days later;

e. The final award shall be announced publicly by the Authority within seven days after the bids have been opened.

3.6. In the event of the termination, forfeiture, or revocation of an exploitation license to a block, or relinquishment of a part of a block, the block or portion thereof will be offered for sale by sealed competitive bidding on a cash bonus basis in addition to the current license fee. The following provisions shall apply to such a sale:

a. The availability of such a block, or portion thereof, for bidding shall be publicly announced by the Authority as soon as possible after it becomes available, and a sale following the above procedures shall be held within one hundred and eighty days after a request for an exploitation license on the block has been received;

b. The bidding shall be opened to all sponsored operators, including, except in the case of revocation, the operator who previously held the exploitation license to the block or to the available portion thereof;

c. If the winning bid is submitted by an operator who previously held the exploitation right to the same block, or to the same portion thereof, the work requirement will begin at the level that would have applied if the operator had continuously held the block.

3.7. Blocks, or portions thereof, contiguous to a block on which production has begun shall also be sold by sealed competitive bidding under the terms specified in paragraph 3.6.

3.8. Blocks, or separate portions thereof, from which hydrocarbons or other fluids are being drained, or are believed to be drained, by production from another block shall be offered for sale by sealed competitive bidding under the terms specified in paragraph 3.6 at the initiative of the Authority.

3.9. Geologic and other data concerning blocks, or portions thereof, open for bidding pursuant to paragraphs 3.6 through 3.8, which are no longer confidential, shall be made available to the public prior to the bidding date. Data on blocks, or separate portions thereof, for which the license has been revoked for violations shall be made available to the public within 30 days after revocation.

3.10. Exploitation licenses shall only be transferable with the approval of the Sponsoring Party and the Authority, provided that the transferee meets the requirements of this Convention, is sponsored by a Contracting Party, and a transfer fee is paid to the Authority in the amount of $250,000. This fee shall not apply in transfers between parts of the same operating enterprise.

4. Duration of Exploitation Licenses:

4.1. If commercial production has been achieved within fifteen years after the license has been issued, the exploitation license shall be extended automatically for twenty additional years from the date commercial production has commenced.

4.2. At the completion of the twenty-year production period referred to in paragraph 4.1, the operator with the approval of the Sponsoring Party shall have the option to renew his license for another twenty years at the rental fees and payment rates in effect at the time of renewal.

4.3. At the end of the forty-year term, or earlier if the license is voluntarily relinquished or expires pursuant to paragraph 5.9 of Appendix A, the block or blocks, or separate portions of blocks, to which the license applied shall be offered for sale by competitive bidding on a cash bonus basis. The previous licensee shall have no preferential right to such block, or separate portion thereof.

5. Work Requirements:

5.1. The annual work requirement fee per block shall be specified in an Annex in accordance with the following schedule:

Paragraph 5.1 (a) and (c) minerals

Years:	Amounts per annum	
1 to 5	$20,000 to	$60,000
6 to 10	$180,000 to	$540,000
11 to 15	$200,000 to	$600,000
Total	$2,000,000 to $6,000,000	

Paragraph 5.1 (b) minerals

Years:	Amounts per annum	
1 to 2	$20,000 to	$60,000
3 to 10	$120,000 to	$360,000
11 to 15	$200,000 to	$600,000
Total	$2,000,000 to $6,000,000	

The minimum annual work requirement fee for a portion of a block shall be an appropriate fraction of the above, to be specified in an Annex.

5.2. Work expenditures with respect to one or more blocks may be considered as meeting the aggregate work requirements on a group of blocks originally licensed in the same year, to the same operator, in the same category, provided that the number of such blocks shall not exceed sixteen in the case of the category of minerals described in paragraph 5.1(a) of Appendix A, four in the case of the category of minerals described in paragraph 5.1(b) of Appendix A and eight in the case of the category of minerals described in paragraph 5.1(c) of Appendix A.

5.3. Should the aggregate work requirement fee of $2,000,000 to $6,000,000 be spent prior to the end of the thirteenth year, an additional work requirement fee of $25,000–$50,000, as specified in an Annex, shall be met until commercial production begins or until expiration of the fifteen-year period.

5.4. After commercial production begins, the operator shall at the beginning of each year deposit $100,000 to $200,000 as specified in an Annex, or with the Sponsoring Party post a bond for that amount. Such deposit or bond shall be returned in an amount equivalent to expenditures on or related to the block and the value of production at the site. A portion [a figure between 50% and 66⅔% will be inserted here] of any funds not returned shall be transmitted to the Authority.

6. Unit Management:

The Operations Commission shall assure unit management and production pursuant to Section 9 of Appendix A, giving the licensees and their Sponsoring Parties a reasonable time to reach agreement on a plan for unit operation.

Appendix C:
Terms and Procedures for Licenses in the International Trusteeship Area

1. General:

1.1. Unless otherwise specified in this Convention, all provisions of this Convention except those in Appendix B shall apply to the International Trusteeship Area.

2. Entities Entitled to Obtain Licenses:

2.1. The Trustee Party, pursuant to Chapter III, shall have the exclusive right, in its discretion, to approve or disapprove applications for exploration and exploitation licenses.

3. Exploration and Exploitation Licenses:

3.1. The Trustee Party may use any system for issuing and allocating exploration and exploitation licenses.
3.2. Copies of licenses issued shall be forwarded to the Authority.

4. Categories and Size of Blocks:

4.1. The Trustee Party may license separately one or more related minerals of the categories listed in paragraph 5.1 of Appendix A.
4.2. The Trustee Party may establish the size of the block for which exploitation licenses are issued within the maximum limits specified in Appendix A.

5. Duration of Exploitation Licenses:

5.1. The Trustee Party may establish the term of the exploitation license and the conditions, if any, under which it may be renewed, provided that its continuance after the first 15 years is contingent upon the achievement of commercial production.

6. Work Requirements:

6.1. The Trustee Party may set the work requirements at or above those specified in Appendix A and put these in terms of work to be done rather than funds to be expended.

7. Unit Management:

7.1. When a deposit most rationally extracted under unit management lies wholly within the International Trusteeship Area, or astride its landward boundary, the Trustee Party concerned shall assure unit management and production pursuant to Section 9.1 of Appendix A, and shall submit the plan for unit operation to the Operations Commission.

7.2. With respect to deposits lying astride a boundary between two Trustee Parties in the International Trusteeship Area, such Parties shall agree on a plan to assure unit management and production, and shall submit the operation plan to the Operations Commission.

8. Proration:

8.1. The Trustee Party may establish proration, to the extent permitted by its domestic law.

9. Payments:

9.1. Pursuant to Subparagraph (e) of Article 28, the Trustee Party may collect fees and payments related to the issuance or retention of a license in addition to those specified in this Convention, including but not limited to payments on production higher than those required by this Convention.

9.2. The Trustee Party shall transfer to the Authority a portion [a figure between 50% and 66⅔% will be inserted here] of the fees and payments referred to in paragraph 9.1 except as otherwise provided in paragraphs 3.3, 4.4 and 6.4 of Appendix A.*

10. Standards:

10.1. The Trustee Party may impose higher operating, conservation, pollution, and safety standards than those established by the Authority, and may impose additional sanctions in case of violations of applicable standards.

11. Revocation:

11.1. The Trustee Party may suspend or revoke licenses for violation of this Convention, or of the rules it has established pursuant thereto, or in accordance with the terms of the license.

---

*NOTE.— Further study is required on the means to assure equitable application of the principle contained in paragraph 9.2 to socialist and non-socialist parties and their operations.

Appendix D:
Division of Revenue

1. Disbursements:

1.1. All disbursements shall be made out of the net income of the Authority, except as otherwise provided in paragraph 2 of Article 74.

2. Administrative Expenses of the International Seabed Resource Authority:

2.1. The Council, in submitting the proposed budget to the Assembly, shall specify what proportion of the revenues of the Authority shall be used for the payment of the administrative expenses of the Authority.

2.2. Upon approval of the budget by the Assembly, the Secretary-General is authorized to use the sums allotted in the budget for the expenses specified therein.

3. Distribution of the Net Income of the Authority:

3.1. The net income, after administrative expenses, of the Authority shall be used to promote the economic advancement of developing States Parties to this Convention and for the purposes specified in paragraph 2 of Article 5 and in other Articles of this Convention.

3.2. The portion to be devoted to economic advancement of developing States Parties to this Convention shall be divided among the following international development organizations as follows:*

3.3. The Council shall submit to the Assembly proposals for the allocation of the income of the Authority within the limits prescribed by this Appendix.

3.4. Upon approval of the allocation by the Assembly, the Secretary-General is authorized to distribute the funds.

Appendix E:
Designated Members of the Council

1. Those six Contracting Parties which are both developed States and have the highest gross national product shall be considered as the six most industrially advanced Contracting Parties.

*NOTE.—A list of international and regional development oranizations should be included here, indicating percentages assigned to each organization.

2. The six most industrially advanced Contracting Parties at the time of the entry into force of this Convention shall be deemed to be:

—————————————————.

They shall hold office until replaced in accordance with this Appendix.

3. The Council, prior to every regular session of the Assembly, shall decide which are the six most industrially advanced Contracting Parties. It shall make rules to ensure that all questions relating to the determination of such Contracting Parties are considered by an impartial committee before being decided upon by the Council.

4. The Council shall report its decision to the Assembly, together with the recommendations of the impartial committee.

5. Any replacements of the designated members of the Council shall take effect on the day following the last day of the Assembly to which such a report is made.

Appendix 18

Draft Articles on the Breadth of the Territorial Sea, Straits, and Fisheries Submitted by the United States August 3, 1971*

Article I

1. Each State shall have the right, subject to the provisions of Article II, to establish the breadth of its territorial sea within limits of no more than 12 nautical miles, measured in accordance with the provisions of the 1958 Geneva Convention on the Territorial Sea and Contiguous Zone.

2. In instances where the breadth of the territorial sea of a State is less than 12 nautical miles, such State may establish a fisheries zone contiguous to its territorial sea provided, however, that the total breadth of the territorial sea and fisheries zone shall not exceed 12 nautical miles. Such State may exercise within such a zone the same rights in respect to fisheries as it has in its territorial sea.

Article II

1. In straits used for international navigation between one part of the high seas and another part of the high seas or the territorial sea of a foreign State, all ships and aircraft in transit shall enjoy the same freedom of navigation and overflight, for the purpose of transit through and over such straits, as they have on the high seas. Coastal States may designate corridors suitable for transit by all ships and aircraft through and over such straits. In the case of straits where particular channels of navigation are customarily employed by ships in transit, the corridors, so far as ships are concerned, shall include such channels.

2. The provisions of this Article shall not affect conventions or other international agreements already in force specifically relating to particular straits.

* U.N. doc. A/AC.138/SC.II/L4.

Article III

1. The fisheries and other living resources of the high seas shall be regulated by appropriate international (including regional) fisheries organizations established or to be established for this purpose in which the coastal State and any other State whose nationals or vessels exploit or desire to exploit a regulated species have an equal right to participate without discrimination. No State Party whose nationals or vessels exploit a regulated species may refuse to cooperate with such organizations. Regulations of such organizations pursuant to the principles set forth in paragraph 2 of this Article shall apply to all vessels fishing the regulated species regardless of their nationality.

2. In order to assure the conservation and equitable allocation of the fisheries and other living resources of the high seas, the following principles shall be applied by the organizations referred to in paragraph 1:

A. Conservation measures shall be adopted that do not discriminate in form or in fact against any fishermen. For this purpose, the allowable catch shall be determined, on the basis of the best evidence available, at a level which is designed to maintain the maximum sustainable yield or restore it as soon as practicable, taking into account relevant environmental and economic factors.

B. Scientific information, catch and effort statistics, and other relevant data shall be contributed and exchanged on a regular basis.

C. The percentage of the allowable catch of a stock in any area of the high seas adjacent to a coastal State that can be harvested by that State shall be allocated annually to it. The provisions of this sub-paragraph shall not apply to a highly migratory oceanic stock identified in Appendix A.*

D. The percentage of the allowable catch of an anadromous stock that can be harvested by the State in whose fresh waters it spawns shall be allocated annually to that State.

E. With respect to sub-paragraphs C and D above:

(1) [The percentage of the allowable catch of a stock traditionally taken by the fishermen of other States shall not be allocated to the coastal State. This provision does not apply to any new fishing or expansion of existing fishing by other States that occurs after this Convention enters into force for the coastal State.]†

---

* Appendix A is not attached. [Footnote in original.]

† It is the view of the United States Government that an appropriate text with respect to traditional fishing should be negotiated between coastal and distant water fishing states. [Footnote in original.]

(2) The allocation to the coastal State shall not be implemented in a manner that discriminates in form or in fact between the fishermen of other States.

(3) When more than one coastal State qualifies for an allocation of a percentage of a stock, the total amount which may be allocated shall be equitably divided in accordance with principles of this Article.

F. All States including the coastal State may fish on the high seas for that percentage of the allowable catch not allocated in accordance with this Article.

3. The provisions of paragraph 1 shall not apply in the event that States directly concerned, including the coastal State, are unable or deem it unnecessary to establish an international or regional organization in accordance with that paragraph for the time being. In that event:

A. In the case of a highly migratory oceanic stock identified in Appendix A, such stock shall be regulated pursuant to agreement or consultation among the States concerned with the conservation and harvesting of the stock.

B. In the case of any other stock, a coastal State may implement the principles of paragraph 2 provided:

(1) The coastal State has submitted to all affected States its proposal for the establishment pursuant to paragraph 1 of an international or regional fisheries organization applying the principles of paragraph 2;

(2) Negotiations with other States affected have failed to produce, within four months, agreement on measures to be taken either with respect to the establishment of an organization or with respect to the fisheries problems involved;

(3) The coastal State has submitted to all affected States the available data supporting its measures and the reasons for its actions.

The implementing regulations of the coastal State may apply in any area of the high seas adjacent to its coast or, with respect to an anadromous stock that spawns in its fresh waters, throughout its migratory range.

4. Enforcement of the fisheries regulations adopted pursuant to this Article shall be effected as follows:

A. Each State Party shall make it an offense for its nationals and vessels to violate the fishery regulations adopted pursuant to this Article.

B. Officials of the appropriate fisheries organization, or of any State so authorized by the organization, may enforce the fishery regulations adopted pursuant to this Article with respect to any vessel fishing a regulated stock. In the event an organization has not been established in accordance with this Article, properly authorized officials of the coastal State may so enforce these regulations. Actions under this sub-paragraph shall be limited to inspection and arrest of vessels and shall be taken in such a way as to minimize interference with fishing activities and other activities in the marine environment.

C. An arrested vessel shall be delivered promptly to the duly authorized officials of the State of nationality. Only the State of nationality of the offending vessel shall have jurisdiction to try any case or impose any penalties regarding the violation of fishery regulations adopted pursuant to this Article. Such State has the responsibility of notifying the enforcing organization or State within a period of six months of the disposition of the case.

5. The international or regional fisheries organizations referred to in this Article shall, *inter alia,* promote:

A. Cooperation with the United Nations, its specialized agencies and other international organizations concerned with the marine environment;

B. Scientific research regarding fisheries and other living resources of the high seas;

C. Development of coastal and distant water fishing industries in developing countries.

6. Exploitation of the living resources of the high seas shall be conducted with reasonable regard for other activities in the marine environment.

7. Any dispute which may arise between States under this Article shall, at the request of any of the parties, be submitted to a special commission of five members, unless the parties agree to seek a solution by another method of peaceful settlement, as provided for in Article 33 of the Charter of the United Nations. The commission shall proceed in accordance with the following provisions:

A. The members of the commission, one of whom shall be designated as chairman, shall be named by agreement between the States in dispute within two months of the request for settlement in accordance with the provisions of this Article. Failing agreement they shall, upon the request of any State Party, be named by the Secretary-General of the United Nations, within a further two month period, in consultation with the States in dispute and with the President of the International Court of Justice and the Director-General of the Food and Agriculture Organization of the United

Nations, from amongst well-qualified persons being nationals of States not involved in the dispute and specializing in legal, administrative or scientific questions relating to fisheries, depending upon the nature of the dispute to be settled. Any vacancy arising after the original appointment shall be filled in the same manner as provided for the initial selection.

B. Any State Party to proceedings under these Articles shall have the right to name one of its nationals to sit with the special commission, with the right to participate fully in the proceedings on the same footing as a member of the commission but without the right to vote or to take part in the writing of the commission's decision.

C. The commission shall determine its own procedure, assuring each party to the proceedings a full opportunity to be heard and to present its case. It shall also determine how the costs and expenses shall be divided between the parties to the dispute, failing agreement by the parties on this matter.

D. The special commission may decide that pending its award, the measures in dispute shall not be applied.

E. The special commission shall render its decision, which shall be binding upon the parties, within a period of five months from the time it is appointed unless it decides, in case of necessity, to extend the time limit for a period not exceeding two months.

F. The special commission shall, in reaching its decisions, adhere to this Article and to any agreements between the disputing parties implementing this Article.

G. Decisions of the commission shall be by majority vote.

8. The provisions of this Article shall not affect conventions or other international agreements already in force specifically relating to particular fisheries.

Index

Index

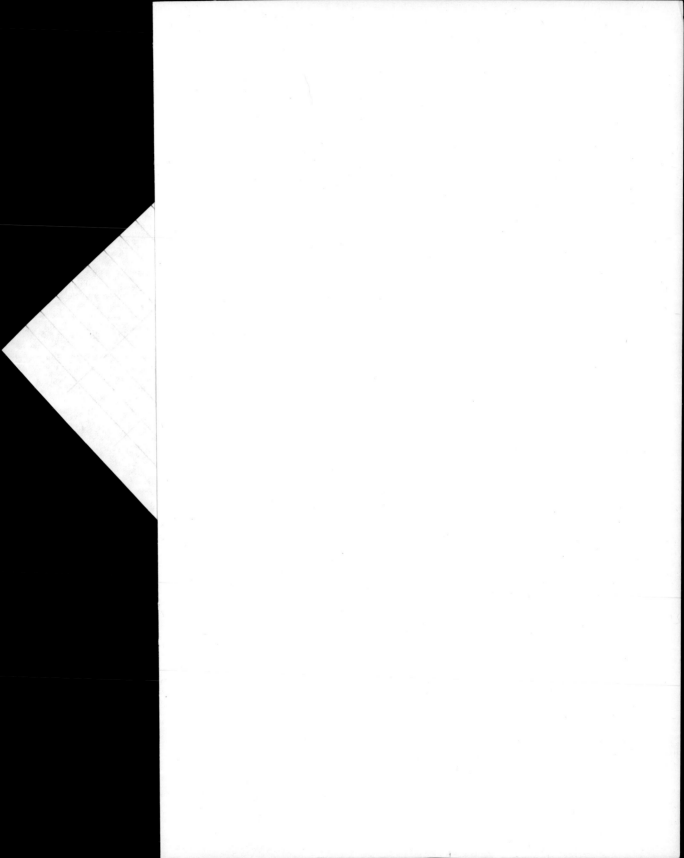